I0086145

How to Survive the Next 100 Years

Lessons from Nature

Simon Mustoe

wildiaries

www.simonmustoe.blog

wildiaries

First published 2025 by Wildiaries
Melbourne, Australia.

ISBN 978 0 6454 5358 4 (Trade Paperback)
ISBN 978 1 7638 1300 7 (Hardback)

© Simon Mustoe, 2025
Internal Illustrations © Simon Mustoe, 2025
Book design, layout and cover art by Nada Backovic
Cover images by Unsplash, Shutterstock, iStock
and Adobe Stock

This work is copyright. Apart from any use permitted
under the Copyright Act 1968, no part of this publication
may be reproduced by any process, stored in a retrieval
system, or transmitted, in any form, or by any means
(electronic, mechanical, photocopying, recording or
otherwise), including being reproduced or used in any
manner for the purpose of training artificial intelligence
technologies or systems, nor may any other exclusive right
be exercised, without the permission of the publisher.

To Carla.

wildiaries-travel.com
Expert led small group expeditions
to the Coral Triangle and beyond.

This sun bear has climbed high above the ground, attracted by the sweet smell of fruiting figs. The bear's behaviour has shaped the forest, influencing its structure for perhaps millions of years. This has formed patterns and relationships with other creatures, such as the stingless fig wasps that swarm around its favourite food. The destiny of fruits, wasps and bears were intertwined until they became the ecosystem on which human civilisation was built. Our future might be in the paws of this creature and many like it. We can learn from sun bears and discover how to behave, transforming our society by putting nature at its centre.

Contents

Foreword

Humankind is transforming the planet through natural resources exploitation and for food production, housing and infrastructure. All this is being done in the interest of enhancing human economic welfare and wellbeing.

But this is also causing a lamentable decline in living spaces available for the millions of other species with which we share the planet. This is especially true for animal species that are suffering a disproportionate brunt of the change.

In this wonderful book, Simon Mustoe tells the stories of how many of those animal species provide vital environmental services that ultimately sustain nature and human economic welfare and wellbeing together.

In a highly engaging and accessible narrative, the book provides a wonderful and broadly encompassing scientific portrait about why we should care about and act to prevent the loss of these amazing and vital members of our shared planet.

Oswald Schmitz, Professor of Ecology, Yale School of the Environment

Introduction

As animals our brains float above the planet's surface; there is an air gap between us and the Earth. This makes us unique among life forms and quite distinct from plants because we were made to be mobile and carry our intelligence with us. Contained within our brains is the reason animals like us can find shelter, food and water.

But it is the overarching systems we live within – those that exist outside our minds – that enable us to have access to the materials for life. The everyday choices we make as individuals aren't as independent as we would like to think they are. Our minds are forced to behave in certain ways by factors both inside and outside our bodies that we don't control. Ultimately, it is these actions that connect us to the planet's life support structures and determine whether our population grows or declines. These extraordinary global structures that our brains float above and our bodies need to survive are called ecosystems. They are built by a huge labour force of animals, of which humans are only one small part.

By working together all animals gather and store up-to-date knowledge to move, act and make democratic choices for ecosystem balance. But because other animals can't talk to us in words, it's imperative we learn to understand them in other ways so that we can act wisely on their behalf. We must ensure we give them the autonomy they need to serve our best interests. To do this we must take our lessons from nature.

In the *Living Planet Report*[1], 70 per cent of all individual wild animals on Earth were said to have disappeared in the last 50 years. The data has since been reanalysed and on balance it appears as though wildlife numbers haven't increased or decreased overall. After researchers excluded a small percentage of the most extreme declines about 50 per cent of species were found to be increasing in number[2].

Scientists recently identified just 20 animals – including bison, tiger, reindeer, black bear, beaver and hippopotamus – that could save half of the world's landscapes[3]. The opening paragraph of the abstract reads:

Assemblages of large mammal species play a disproportionate role in the structure and composition of natural habitats. Loss of these assemblages destabilizes natural systems, while their recovery can restore ecological integrity.

Not only does this place attention on the restoration of important systems but recognises that animals themselves build ecosystems, as we will discover later in this book. *The Living Planet Report* is one of the first papers of its kind and represents a shift in the international research psyche from one where animals are considered the icing on the cake of ecosystems, towards one where their conservation is considered essential to the very substance of ecosystems and, therefore, our survival.

Successful wildlife recovery, even of large animals, can happen fast. By the year 2000, saiga antelope numbers had collapsed by 95 per cent to 15,000 animals because of illegal poaching. Today, thanks largely to the Altyn Dala Conservation Initiative working with local people and the Kazakhstan's government, they are better protected and now number 842,000. In Europe's landscapes brown bears, wolves and lynx are recovering faster than almost anywhere else on Earth[4].

Author, evolutionary biologist and film-maker Sean B. Carroll made the Oscar-winning HBO nature documentary *All That Breathes*. It tells the story of our connection to and dependence on nature's ability to clean up our mess through the wisdom of two brothers saving kites on the streets of Delhi. He believes we aren't doing enough to tell the story about nature's resilience. The film contains the quote 'man is the loneliest animal', which is what we become when separated from wildlife. But he believes wildlife can rebuild our world dramatically, perhaps in as little as 20 years, when given the chance[5].

We need wildlife in the balance, in the right proportions and in abundance. Where we've lost a lot of the bigger animals, we have made other species inconveniently abundant. Nature is readjusting and it's closing in on us because collapse is the natural order of the Universe and stable ecosystems only form when animals regroup into structures that are the most resistant to decline. Think of ecosystems like pyramidal sand dunes that last thousands of years rather than being blown into the sea by desert winds.

Ecosystems are constantly blown apart by the Sun's energy. It breaks down their structure before they are rebuilt by animals that form pyramids of life, starting with a few predators at the top and finishing with innumerable microorganisms at the base. For our own benefit, what we need most of all is a change in human values where we recognise the importance of all wildlife in balancing ecosystems and protecting our collective future.

Fifty years of environmental change might define a civilisation for us, but it is nothing to nature. We are only at the beginning of a cycle of thousands of years of learning new lessons. Nature has always been there to nurture us through the hard times. We're already adjusting on a huge scale as the rebalancing of our planet progresses through climate, ecosystems and wildlife habits. It is critical that we give remaining animals the autonomy to move around and restore their own cultural connection, to maintain their civilisations while we figure out our own through our relationship to them and to nature. As we will discover throughout this book nature teaches us that as we yield to it and learn to connect with the wildlife around us, together we can quite quickly rebuild a habitable world.

Nature is everything when it comes to considering our future as a species – our very civilisation has always depended on it. As we will learn in 'Lessons from our Past and Present', through our shared similarity with wildlife we can find examples in the history of our own species and that of other animals that offer clues to how we've survived in the past and will, most likely, learn to survive again in the future.

'Lessons from Our Connection to Creatures' explores some of the strange yet common ways nature compels us to act, think and behave. Given that the complex web of life that we are part of can only really be interpreted through our relationship with other animals, it stands to reason that they are the medium through which we learn the most valuable of lessons.

As with 'sustainability' and 'biodiversity', 'nature' is too complex a concept to be given a simple definition. When we try to pigeonhole it or create a unifying approach to addressing it, we fail. There are no simple, right or wrong definitions because everyone's connection to nature differs in time and place. It is best to see it as a cause to believe in (like 'health' or 'nutrition') and take actions to protect and enhance it. To have a fruitful life we must strive to restore nature and biodiversity and enact sustainability.

The lessons we learn about nature's prowess and our connection to it are plain to see if we take the time to look and understand. From the sea creatures we explore in 'Lessons from the Ocean' to the more familiar animals we are accustomed to living alongside in 'Lessons from the Land', animals have a monumental and positive impact on our survival. 'Lessons from the Air' explores some of the ways animals have dominated the skies – threading through our world's wetlands, maintaining water cycles and creating an overall peaceful and habitable climate.

All of this culminates in the stabilisation of our climate, and as a result it is in all these lessons from nature that we find the best hope of survival for humanity.

For those who know how to look there are wonderous and rapidly multiplying examples of where people and nature are uniting on a mass scale to make a difference to both our individual lives and our collective future. This book isn't about hope for the mere sake of it, instead it is an attempt to reset the imbalance that permeates our culture and biases our world view, making us imagine all hope is lost. Those narratives aren't true. As we will discover, new money, old beliefs and modern science are already working together to ignite change in a most spectacular way.

Changes are happening everywhere in all walks of life when it comes to locally initiated ecosystem recovery from large financial and insurance companies reporting their biodiversity losses and investing in nature recovery, to local people running petitions to save the nature in their own backyard or creating community gardens. Many animals are making a comeback themselves with little help at all.

There are celebrated conservation programs for many threatened species – most notably animals like orangutans, elephants and koalas – and widespread reintroductions of species like European lynx, bison and beavers. But on top of any human-led initiatives is the natural recovery of animals like humpback whales. In addition to this, the re-emergence of iconic species into humanity's consciousness through media and tourism continues to change hearts and minds when it comes to conserving our natural world.

When all this evidence is added together in 'Lessons from Giants' it tells a different story. The one we're used to hearing is of a dying world filled with doomed species – our own included. Instead it is a tale of our world rapidly changing for the better.

Imagine the outcome if we all knew of, and could become involved in, these extraordinary interventions. The promise of hope – rather than the relentless cycle of disaster that we've become accustomed to around nature, climate and animals – would impact our collective consciousness to the extent that people might instantly become happier and more hospitable to each other. Hope and purpose are catalysts for cooperation that merge to become belief in a more optimistic future and when we cooperate conflict quickly dissolves. This more harmonious existence all begins with you, dear reader.

A tidal wave of influence is occurring, motivated by a deep-seated and primaeval need to save not only our species but also our planet and everything that inhabits it. In my 50 years on Earth, I have never witnessed so much wholesale and widespread positive action and concern for nature as I am now. People all over the world are beginning to accept and understand nature, to know why they need healthy ecosystems.

In some cases, humans are even adopting customary rights for its protection. Ecuador, for example, was the first country to recognise the inalienable rights of nature; Chapter 7 of its constitution was redrafted in 2007, before a referendum where over 80 per cent of its people voted in favour of the change. It now reads:

> Nature, or Pacha Mama, where life is reproduced and occurs, has the right to integral respect for its existence and for the maintenance and regeneration of its life cycles, structure, functions and evolutionary processes … all persons, communities, peoples and nations can call upon public authorities to enforce the rights of nature … Persons, communities, peoples, and nations shall have the right to benefit from the environment and the natural wealth enabling them to enjoy the good way of living.

Since then, more than 35 countries have followed Ecuador's lead in some form or other: Bangladesh granted legal personhood to the Turāg River; New Zealand has done the same for the Whanganui River and Mount Taranaki volcano; and the Mar Menor salt lagoon in Spain is the first European example.

Elsewhere, regional councils are adopting their own charters to recognise the life support that nature provides: the Blue Mountains City Council was among the first in Australia; Donegal in Ireland has done similarly; and the Brazilian city of Linhares has just given rights to the Doce River mouth to protect surfing waves. Communities and farms throughout the world – their businesses already given personhood in law – are also planning a return to nature-based living and nurturing wildlife back into the environment.

We are also beginning to embrace our indigenous cultural and natural history knowledge to protect and restore nature's assets, helping get back what we've lost from degrading it for so long. Globally, some the world's greatest economic influencers have realised this is key to unlocking financial benefits bigger than anything we've been used to in recent years.

By better connecting people with wildlife, animals will have the space to restore ecosystems and we can turn the hopes and aspirations of billions of people into actual change that will greatly improve all our futures. Nature is teaching us these valuable lessons already. This action hasn't come about because someone told us we must do this; it has come about because it's a necessity if we want life as we know it to continue. The campaign for our survival has risen from the ashes of dying ecosystems and is escalating bewilderingly fast. We are being forced to redefine our relationship with nature and wildlife so that we can remedy cost of living pressures and secure our livelihoods every day. The belief in rights for nature stems from a stark economic need to live a more comfortable existence and is becoming the most important step towards humanity's survival in the next 100 years.

Perhaps the most extraordinary change of all is how animals are altering their own behaviour and are being forced to appear among us where they once hid away. Even species we thought we'd lost are coming back and rebuilding ecosystems in places where it's no longer economically viable for humans to pose a threat to these animal-driven processes[6]. This is a phenomenal change you won't read about anywhere else because it's obscure and can't be explained fully by written history or modern science. The knowledge that this is true rests in most of our minds and is inherent in our age-old wisdom as creatures of the natural world.

Many of our greatest technological advances are inconsequential compared to the advances we are making by reconnecting and rebuilding with wildlife. We are approaching a turning point where things will start to get better again for our species and the living planet, where the return of wildlife will bring improvements to our lifestyles and livelihoods that we can currently only dream about.

Humans aren't about to become extinct anytime soon – it's only our egos that make us think we can destroy everything. As much as we like to think it, we're not that powerful. Nature is.

The ways we connect to wildlife and nature are infinitely weird and wonderful, but they are also a reminder that we don't act alone in this world. We have little control of our destiny, except insofar as we are a functioning part of the wildlife-driven ecosystems that we live in. The reality of our existence is much simpler than we imagine. We survive because of an abundance of wildlife in the balance. All we need do is keep telling ourselves this simple truth and we will find proper cause to hope.

But first, there are important lessons to learn from nature. Lessons we need to learn before we can build a healthy respect for the world around us and survive the next 100 years.

Lessons from our Past and Present

Given three choices, how wisely do you think you could make an informed decision to improve your chances of survival?

In 2014 the scholar Hans Rosling talked about an experiment where he had asked thousands of people around the world a series of multiple-choice questions about the state of our society. Rosling wanted to consider how long it takes for societies to catch up with facts that were learned many years before. One question that formed part of this experiment was: how did the percentage of people living in extreme poverty (those without enough for food for a day) change in the last 20 years?

Of the participants from Sweden, 50 per cent thought it had doubled, 38 per cent said unchanged and 12 per cent said it had declined. Rosling then did something that might seem strange – he went to the zoo and asked the caged chimpanzees. The chimps chose answers randomly because, as Rosling noted, they 'don't watch the evening news'[7].

As it turns out, poverty has halved in the last 20 years. Strangely enough the estimate by the chimps was the closest and this ended up being the case for most of the questions Rosling asked. When participants in the US were asked the same question, only five per cent got it right.

That is how much influence culture has on your lives and livelihoods. What you think you know determines what you believe, and if you are told that we are all doomed by the cultural influences that permeate our daily lives the likelihood is you're going to believe it.

Rosling's experiment was a little unfair on chimpanzees given that being an expert on their own ecosystem is literally life and death to them. If you could ask a wild, locally native chimpanzee where to find fruit, it would tell you with precision by scanning the forest for trees that fruit at similar times. It would use its botanical knowledge combined with the understanding of seasons and acute navigational skills learned by its ancestors and passed down over generations.

But we could never ask a chimpanzee whether the extinction of a fruiting tree is more important than the life of one chimpanzee. In all the years of scientific experiments asking chimpanzees to answer questions and perform cognitive tests in labs, not one chimp has ever asked any human a question. It's not in their nature. They are just behaving in the way that best ensures their species' survival – as their kind has done for millions of years.

A chimpanzee's whole society depends on certain tree species surviving and chimpanzee culture is bonded to that knowledge. But change the ecosystem, threaten the existence of that tree and its culture becomes corrupted. Corrupt its culture and knowledge and chimpanzee society collapses. Put a chimpanzee in a cage away from its natural habitat and its choices become random, no longer defined by the world it is meant to exist in. Subject it to the news and its choices become imprecise, misleading and biased. It's a pointless question to ask a chimpanzee and it is, therefore, even more pointless to ask a human who watches the news if nature is supremely important. We are too disconnected from nature to make an unbiased assessment.

Western culture is easily swayed or corrupted, meaning that – if not properly informed – social bias can lead its population off a cliff. The danger that humanity faces lies in the point where this social bias starts to affect the delicately balanced natural world around us. There are examples of this everywhere; one recent study found that Australians would prefer actions that could save a single human life from a bushfire, even if this meant driving an entire non-human species to extinction[8].

In contrast, Australia's long surviving First Nations culture is based on unmitigated respect for the environment – an unbreakable bond between people and nature that even outlasted 100 years or more of atrocious cultural oppression. Human knowledge systems evolved over many thousands of years to become a blueprint for the long-term survival of civilisations. Today there are still profound examples of stories that have retained their clarity and precision for longer than any others in human history. It is the most sophisticated knowledge we have about how to survive and yet it is often overlooked when we are seeking answers on how to protect our future. This is wisdom in its truest form.

This ancient wisdom was enabled by a different type of scientific thought than the one we depend on today and is particularly founded on the strong belief that we are beholden to nature and wildlife. It was built on trial and error, life and death, conversation and consensus just like the wisdom of many other species in our world. The majority survived because a balance was achieved between humans and wildlife – a strong belief in the importance of nature and culture held this race intact. By understanding nature and connecting with the world around us, we learned to make good decisions for our environment and social welfare without having to ask.

When we disconnect from both nature and our past wisdom, our culture suffers. We are no longer able to make wise decisions for our own welfare. That we are dependent on nature is both a unifying, age-old truth and the greatest lesson of all that we can learn from wildlife.

The Budj Bim

Over Christmas 2023 we visited the Budj Bim cultural centre in western Victoria, Australia as we were intrigued to find out why somewhere most of us have probably never heard of was declared a World Heritage Area in 2019. This ancient landscape joins a prestigious list of other heritage sites you definitely will have heard of such as Machu Picchu, the Pyramids, Stonehenge, the Great Barrier Reef, the Taj Mahal and the Great Wall of China.

While all these locations are globally recognised for their universal heritage values, this one is different. Many of the others are archaeological monuments or wilderness areas, but Budj Bim is a living, breathing and functioning landscape that has not only supported human life for thousands of years but continues to do so today. How? Because this place is an example of sustainable aquaculture on a scale that modern people can only dream about.

At a time when the world is struggling to rebuild farming systems, the wisdom of the Gunditjmara nation is making an important contribution to how we can use knowledge and practices from the past to rebuild our future through ecosystems like this one. An entire nation, their nation, has survived for tens of thousands of years by amplifying the positive impact of short-finned eel ecology on the landscape and farming it for food. In working with a super-abundant migratory animal, the Gunditjmara were able to turn a landscape broken by a volcanic eruption into a place where humans could thrive thanks to the consequential wealth of wildlife that concentrated around this activity.

On visiting Budj Bim the first thing you notice is the sheer number of birds, even though this is only a fraction of what would have existed before European settlement 250 years ago. Today Australian bustards are extinct, red-tailed black cockatoos are critically endangered, magpie geese have disappeared and wetland birds like brolgas are much depleted. The Gunditjmara's culture embraces the right of nature to exist, so there are already plans to restore these animals to the lakes and wetlands in future as they know this is what's needed to bring the country to life.

The second thing you notice after passing through the gate into this Gunditjmara-owned land is the health of the vegetation. The landscape is tussocky and diverse, brought back to life in a thriving mosaic of lush green phragmites, dense stands of juncus and basalt-strewn grass plains. In contrast, the nearby Lake Condah Estate floodplain is bowling-green flat, featureless, dusty and barren, fringed by drainage ditches, stripped of moisture and with a few cattle on the horizon. The lake is at the centre of the heritage area but is sadly now only about one-fifth of its original size because water has been tapped upstream and used to irrigate land that is increasingly desertified.

Ironically, systems like this where animals and water flow freely together are essential to the retention and slow release of water, which would have kept surrounding land moist and productive for growing food. In the past, water allocation was responsive and equitable to feed local people, but modern agriculture relies on allocation to the most economically valuable exports and that denies everyone else who lives nearby a 'fair go'. The surrounding land eventually collapses – as it has here – jeopardising the health of everything around it.

People alone don't define an ecosystem's significance; it is always through connection to other creatures that we understand the true importance of these environments. Nowhere is this more obvious than in the example of how the people of the Budj Bim learned to create the world's largest and longest-surviving aquaculture system to farm short-finned eels (called *kooyang* in Dhauwurd Wurrung – the language of the Gunditjmara people).

Australia's second-longest serving Prime Minister, John Howard, famously advocated the forced teaching of European colonial history. Apparently this was so we could learn from past mistakes, but what's the point of it at all when such history is based on deliberately denying the existence of our most significant human endeavours? Where is the teaching on past successes such as Budj Bim that can help create a more balanced ecological future?

Colonial history endures. Australian historian Bill Gammage in his book *The Biggest Estate on Earth*[9] commented about how quickly the deep reds and violet hues of native vegetation became obliterated by sheep and cows when colonists arrived in Australia. Just decades after, artists began depicting grassy Australian landscapes as golden in colour. This impression is as fake as the name 'Mt Eccles' given to the 60 m high dome of the ancient volcano Budj Bim visible just above Lake Condah's tree line.

> The settlers renamed the sacred site of Budj Bim as Mt Eeles in honour of an English aristocrat. Due to a map misprint later in the century, this in turn became the totally meaningless Mt Eccles[10].

That the site was intended to be named Mt Eeles is a fitting coincidence given how much the short-finned eel has come to define this landscape.

What you learn from a half-day tour of Budj Bim is a globally important history of Australia. Though quite different from the romanticised post-colonial lessons taught in school, the reality is far more interesting and relevant to our future. It is a story so old and compelling, and within in it is described what we need to do to imagine and build a better future for all of Australia. Here is our hope for survival, trapped among patterns of boulders rearranged in the landscape of western Victoria.

The dreamtime stories of the Gunditjmara people who built the landscapes of Budj Bim offer an insight into what our future ecosystems could be doing for us. When Budj Bim's domed head spews forth teeth across the landscape it is only the start. The complex relationship between people, wildlife and country that evolved after the volcano cast its teeth (basalt rocks) into the sea became carved into traditions, behaviour, customs, rituals, songs and dances.

Europeans settled in Portland Bay in 1834, about 200 years ago. The Gunditjmara's cultural history is at least 6,000 years old. The most recent uncovering of their advanced land use, trade routes and artefacts through archaeology, bushfires and satellite reveal a civilisation perhaps as large and sophisticated as the city of Cusco during Inca rule. This includes the fishery developed from working with the short-finned eels.

Noel Butlin's estimate of 30–35,000 as the immediate pre-contact First Nations population of the Western District of Victoria indicates the potential number of people involved in the fishery[11]. The dams, ponds and channels engineered by people connect eel ecology and human values to one of the longest-lasting stories in living memory.

The Gunditjmara people's creation stories date back to that event 37,000 years ago when Budj Bim erupted. The ponds and channels that permeate the landscape are like the features of a wizened face, and the enduring relationship between eels and humans provided everything that this thriving society needed to survive longer than any other civilisation on Earth.

The eel defines the region's health and prosperity as without it there would be no towns, no water and no farming today. If it wasn't for the sheer resilience of the Gunditjmara people, our understanding of the significance of this animal's impact on the region could have been lost. We are lucky to have significant fragments of Gunditjmara traditions and stories surviving today alongside the physical landscape they helped build.

There is no doubt that Aboriginal people would have enhanced the population of *kooyong* living throughout this region. In turn, the eels themselves created much of the farmland nutrient on which European-style agriculture was first based. However, in the present day we've exhausted the soil of our land and farming will again become dependent on these animal-driven processes. This is why restoring the methods of Budj Bim is critical to our farming future.

As humans, our connection to the sea is vastly underestimated. The short-finned eels of Gunditjmara country are not the only aquatic wildlife to have an impact on farming. We now know that migrating Northern Hemisphere salmon are responsible for spreading critical nutrients throughout farming landscapes[12]. Salmon, eels and other migratory fish bring nutrients from the deep ocean. Due mostly to a decline in species like eels, 96 per cent of that nutrient transfer capacity has been lost[13]. In other words, eels were laying the foundations for our farming systems long before any humans arrived on the continent.

These short-finned eels mature in the wetlands of southern Victoria before migrating 4,000 km to spawn in the Coral Sea somewhere between Australia and New Caledonia. The juveniles (called elvers) then swim back along the East Australian Current and even climb waterfalls. During summer, their bodies glisten in the moonlight as they abandon the river to find purchase in damp vegetation along its banks. They slowly writhe over the top of each other, inching their way to the summit of seemingly impenetrable barriers such as the 13 metre high Hopkins Falls near Warrnambool. Then they disperse into the region's remaining creeks, lakes and ponds where they will spend a decade or more before reaching full size.

It is here that Gunditjmara people living in the Budj Bim region would have caught a proportion of eels to store and fatten them up in purpose-maintained reed lagoons (the remaining bycatch was returned to the land). Eel numbers at the time would have been vast and this symbiotic relationship of eel and human ecology would have increased landscape fertility exponentially.

Today most rivers, creeks, lakes, dams and swamps on public land south of the Great Dividing Range are open to commercial eel fishing – an activity that extends far inland. Aquatic systems have been shown to be dependent on short-term (decades) and long-term (up to geological timescales) sources of nutrient from eels and other aquatic creatures[14]. Yet these studies don't mention how much that relationship between people and fish would have been a principal and beneficial driver for building land fertility over many thousands of years. Budj Bim shows us why that was possible and how it still can be. By restoring fish populations we can recreate upstream nutrient recycling that farming will come to depend on in future.

Currently we're doing the opposite of that. We 'manage' fisheries as though the resource is only of short-term significance. The Victorian Fisheries Authority is supposed to look after fish numbers, but eels barely register as a concern even though their numbers have plummeted by 95 per cent or more. Fisheries stock assessment ignores the socio-ecological role of animals and turns fishing into an ecological 'sink'; though fish populations can stay measurably unchanged for a long time, the ecosystem beneath inches towards total collapse as there aren't enough animals to maintain the critical underlying processes. These losses mount up, becoming so complex that scientists can't fully understand them. This is why so-called 'tipping points' are a concern. They can happen instantly and occur unexpectedly for reasons we don't see coming, despite research trying to tackle some of the risks.

Once we reach that tipping point everything disappears, including whole farming economies. The vanishing of eels and waterbirds that would have once flourished may already be a sign of that impending ecological collapse. But thanks to the Gunditjmara people we can learn a new lesson from nature – how fast the system is able to recover.

> **LESSON 1:** It is still possible for people and animals to co-exist and build life support ecosystems together.

Fishing, like any extractive process, should be done in the style of the Gunditjmara people of Budj Bim. Often referred to as 'fishing from the top of the barrel' (a nice turn of phrase), this practice promotes sustainability and ecological balance by only taking what is needed. In this process, the barrels (waterways) were kept overflowing and only food required for local sustenance was creamed off the top.

Additionally, the barrel itself is being looked after by the Gunditjmara and the wildlife they associate their culture with. This promotes diverse vegetation growth for water filtration and increases the moisture capacity of soil, all of which leads to more rapid organic processes for food harvesting.

As a bonus, the local climate will become less prone to frequent flash floods. Macro-environmental threats such as decline in soil moisture, stagnation of waterways and the thinning and acidification of soils are reduced. Flourishing wildlife means the landscape adapts faster to natural changes in climate, making it easier to grow food and find clean water. Scientists refer to this as overall increased 'resilience' and it slows the collapse of systems to a crawl. It reduces conflict for resources so cooperation between people and animals increases. To reveal these possibilities all we need to do is remove the threats and enable animals to do the job of rebuilding or maintaining the environment for us. Do less. Spend less. Create more benefit. There are no downsides.

Globally, most fisheries management is a shameful example of how not to manage a landscape. It treats animals as commodities rather than part of a functioning ecosystem, focusing on the individual elements of the environment rather than considering the services they provide as a vital part of the collective. This is fishing from the bottom of the barrel.

The system used by the Gunditjmara people has two channels diverting water from a creek in the lake at Budj Bim. So far, over 1,400 m of these systems have been uncovered. They would have dotted the whole landscape but many were destroyed by farming. Along each of these channels they would have placed a hand-woven cylindrical net to catch eels. In between these culverts are dotted circular pools full of bright green reeds. About 60 per cent of caught eels were returned to the environment, which is enough to maintain life support processes over the remainder of the landscape and ensure the longevity of this life-giving ecosystem. The rest were stored alive in natural pools to grow. Each eel could feed 20–30 people.

To become the source of human nutrition that they were for thousands of years, the eels needed to be farmed in a manner befitting the entire landscape. It's remarkable to think that this was done throughout an ice age and at times when sea levels were 125 m lower than today. The aquaculture and community trading environment were resilient enough to maintain eel migration and feed a resident human population through a huge climatological event – much like the one we are facing today (though ours is in a much shorter timeframe so we must be even more aware of this lesson from nature).

It's thought Budj Bim was one of the most stable and sedentary First Nations societies in Australia because the food was so abundant and reliable they didn't need to be nomadic. They lived in villages and traded as far north as Broken Hill 1,000 km away and, when sea levels were 125 m lower, into Tasmania. The channels they used might have been a metre deep in places, created by heating the rocks then introducing water so that they exploded to form new passages and waterways. It would have been a huge job to create this system, but the rocks are so hard the channels have lasted thousands of years – much like the Gunditjmara's stories and culture.

During our visit to Budj Bim, I asked our guide Reuben about the land's history. 'It was farmed intensively for livestock since the 1850s,' he says. 'We acquired it in 2008 and decided to remove the cattle. The landscape has recovered since then.'

'Recovered' is a modest way of putting it. Fifteen years later and the difference is starkly apparent as red-necked wallabies burst forth from cover habitat and later we're joined by a pair of curious wedge-tailed eagles. These majestic predators are sentinels for the environment's health. Their presence and that of swamp harriers, pods of pelicans, marsh (whiskered) terns, glossy ibis and grazing black-tailed native hens are all indicators of the otherwise hidden potential for an ecosystem coming back to life.

Meanwhile, people argue over water rights to these areas. The once mighty Murray River, navigable all the way to the sea, is lifeless in places. Fish die in their millions and the river dries up. Water is allocated into farm wasteland with sieve-like soil and no way to retain moisture, the majority draining back out to sea or evaporating to re-enter the land-to-sea water cycle and build massive storms that destroy whole towns. It's a waterborne butterfly effect that has enormous consequences – the lost flick of fins and evaporating rivers create a storm in another part of the country.

To survive through these thousands of years the Gunditjmara had to learn to think like a mountain – specifically their mountain, Budj Bim. 'That humans have not learned to think like a mountain is why we have dustbowls and rivers washing the future into the sea,' says Aldo Leopold in the book *Beastly* by Keggie Carew[15]. To think like a mountain is to observe and yield to the steady ebb and flow of everything through an ecosystem – the natural drainage of water around hills and into swamps, the migration and movement of animals up and down river. Water is the substance that makes Earth habitable and in our thirst for profit we have speeded up its flow, making it less available for life support.

According to Erica Gies, author of *Water Always Wins*[16], 'slow water' is the means to combat drought and deluge. This simple philosophy, which might include rebuilding moisture-laden landscapes for eels, can even be used to avoid the increased fire risk over half the world's land areas and half the world's population.

These days the people of Budj Bim must fight to divert water to their land while a tidal wave of freshwater drains or evaporates fast off land elsewhere. At Budj Bim the ecosystem is a sponge able to retain water longest for everyone downstream, slowly releasing it into neighbouring properties. Given that 85 per cent of Australians live on the coast, this is quite important because denying the landscape water today means denying everyone else water forever.

Why, I wonder, are more landowners not screaming advocates for this plan to utilise the knowledge of Budj Bim and revitalise their own waterways? Why aren't they begging to be part of restoration efforts for every creek and wetland throughout the area, rather than argue over the last few drops trickling down the Murray River? The farmers of this region need this water to be used wisely just as much as they need the nutrients the fish bring. What better way than to refill the region's aquifer while also reducing extreme cycles of flood and drought than by rekindling that connection with eels and fish?

The heart of any country is its watercourses, because without water everything dies. The gradual draining and stocking of the remaining land has killed natural human- and animal-driven processes that recycled fertility back onshore for tens of thousands of years, even before the Budj Bim completed their extensive aquaculture system. The remaining river water is now wasted and in doing so loses essential organic nutrients to the ocean without a mechanism to bring them back. Another worrying side effect of this wastefulness is the pollution of coastal reefs and increased coastal erosion.

Today, modern tools like Light Detection and Ranging Lasers (LiDAR) are helping to right historical mistakes by once again revealing the scale and significance of this remarkable place. But in doing this we must also remember that we aren't making discoveries, we are simply rediscovering wisdom that we have allowed to be lost. Some of the associated customs, cultures and harvesting techniques of the Gunditjmara are still in use this century and the people who know these live among us today. It is not the theory but the artefacts that remain hidden below the thin veneer of a post-colonial landscape, flattened and desiccated.

We don't always need science to prove what we already know, but we do need it to help rebuild what we've lost. This means believing in something bigger, something that science will never prove: knowing why the natural world is key and we must respect the delicate balance of humanity, landscape and wildlife if we're to have any chance of surviving the next 100 years.

We cannot manage species; we can only manage our own behaviour and allow animals to look after the land. If we let wildlife populations decline further, we will soon lose all the benefit they can have for our own survival. But in losing wildlife we lose something else, too, in the lessons we learn from nature.

The lesson is that it has always been this way, as the Gunditjmara people will tell you. It is the actions of animals working together that maintains healthy ecosystems. As Reuben says, 'The animals are here to look after the land. We are here to look after the animals.' And here in the Budj Bim World Heritage Area lies a rare place where we can relearn why ecosystems work and how humans can be a force for good.

We need to look to the past and apply the lessons we learn from nature through Budj Bim to the present day. Only a fool would wish to ignore these truths and deny everyone the chance to learn something critically important to all our futures.

Remarkably, despite all but a few words of local Gunditjmara dialect remaining their stories endure because they are as old as time and as firm as the steely basalt rocks that characterise the landscape. Attempts to overpower this wisdom are feeble by contrast. Those who amplify these simple truths and the wisdom they contain will be among the most likely to survive, which is the very definition of natural selection.

Such stories will always prevail because culture and nature are so tightly bound to the landscape that it ultimately means the difference between life and death. This is our first lesson from nature.

Civilised Green Turtles

Green turtles live for 100 years and drag their heavy bodies onto land to lay eggs. They epitomise the slow pace, indefatigable effort and tolerance an animal needs to form a civilisation most likely to survive for millions of years. For this reason, turtles might be among the animals that will probably outlive humanity.

Researchers have found evidence that generations of Mediterranean turtles have been using the same seagrass beds for thousands of years[17]. This is one of the few times scientists have referenced the long-term, ecosystem-linked behaviour of any animals. This significant discovery led me to ponder whether sea turtle civilisation is comparable to human cultures in the Mediterranean, Europe and beyond.

We might equate civilisation with survival, but they aren't the same thing. National Geographic Education defines key components of civilisation to mean 'a complex way of life characterised by urban areas, shared methods of communication, administrative infrastructure, and division of labour'[18]. It is pertinent to note that this is a rather post-colonial definition; I'd imagine First Nations people in Australia would object to any definition of civilisation as only encompassing 'urban areas' given that their civilisation didn't have cities.

The definition of civilisation given above is also inherently linked to humanity, but we are not the only ones capable of building them. Where turtles are concerned, communication and division of labour – the very definition of civilisation – apply. In fact, both are facets of any animal culture that is most likely to survive. Rather than civilisation and survival being synonymous, it becomes apparent that survival is inherently dependent on a different kind of civilisation.

As humans we grossly underestimate the significance of culture, cooperation and the way animals share knowledge with each other. We think wisdom is a purely human trait but it's not. Animals have also learned to make individually wise decisions so their population remains viable.

Some may argue that calling turtle habitat 'civilisation' may be drawing a long bow, but if we consider 'civilised' to mean equitable, long-lasting, stable and occupying one place, then modern sea turtles have enjoyed a very civilised existence. These creatures have been practising for 110 million years. By comparison our civilisations don't survive well at all. All the world's modern, urbanised civilisations of the past have eventually died out.

The longest-lasting known human-built civilisation in Europe was the Minoan, which lasted about 2,000 years. The Romans lasted fewer than 500 years. The Palace of Westminster, which determines what people from the UK would regard as their civilisation of today, has only been the centre of power for about 900 years. The civilisation born out of the colonisation of the United States has lasted about 250 years but who knows how long that will survive.

The oldest animal civilisations tend to last a lot longer than humans and Mediterranean Sea turtles have reportedly managed to exist in one place for at least 3,000 years. The Baltic people practice a language that lifelong orangutan researcher and Lithuanian descendent Dr Biruté Mary Galdikas reminds us is over 8,000 years old and predates Ancient Greek. These indigenous people of Eastern Europe worshipped nature but, like the sea turtles, their nature-based civilisations have been forgotten as we increasingly relate civilisation only to a modern, urban form of living.

It is a trait within all animals individually to constantly destroy and rebuild their environment to suit their personal survival. Green turtles eat seagrass but, more significantly, they contribute to a system that creates more seagrass to replace what they've removed. If they didn't, they would go extinct.

Humanity depends on seagrass, too. Seagrass is essential to all life on Earth and you couldn't be here without it. Seagrass is valuable because it fuels coastal ecosystems, protects us from coastal erosion and is a breeding ground for most of the fish we eat. Commercial fisheries underpin our economies as most people live at or near the coast. It's thought that seagrass absorbs 10 per cent of the world's carbon dioxide, providing us with a trillion dollars of such value globally[19], triple that if you include nutrient cycling, fish habitat and so forth[20].

The impact of animals on the world's ecosystems and the ecological stability that creates culminates in all the basic elements of life – such as food and water – becoming available where and when it befits our survival. Let's call this a 'fair climate' for living. This is a significant lesson to learn because climate is the grand finale of animal impact. It's the ticker-tape parade and festival that signifies the combined successes of all animals working together.

The fact of the matter is that without turtles we can't exist. Why? Because if we remove all the other animals that create seagrass ecosystems, all we as humans can do is destroy them. Seagrass ecosystems are vital to all life on Earth and yet we can't replenish them as we don't have the power to do that. Animals do. They are a necessary component to human civilisation's survival. Why then, doesn't that definition of civilisation also include the need for abundant wildlife? It's almost as though for something to be classified as a civilisation it must also disregard the outside support and influence nature has that is inherently linked to the actions of humanity.

There is a common misconception widely referenced in the media that wildlife plays little or no part in replenishing the ecosystems they are part of. UK television presenter and gardener Alan Titchmarsh from the popular shows *Gardeners' World* and *Ground Force* was quoted in *The Independent* newspaper[21] as saying that 'rewilding does not increase plant diversity'. He adds, 'Wildlife is "adaptable" and can learn quickly what plants are of value, no matter where they come from.'

But Titchmarsh is talking about domestic gardens that often have little to do with ecosystems, and any audience to his comments likely won't realise that distinction. This is where the danger lies in discussing these terms and topics through general media. To put it bluntly, we get the wrong end of the stick and think that theories that apply to domestic gardens also apply to ecosystems.

It is important to understand this distinction. Titchmarsh is implying that animals rely on individual plants, which they do in a small-scale garden but not in larger ecosystems. On a larger scale, humans rely on plant diversity that in turn relies on animals – the opposite of how a garden works. In ecosystems animals increase plant diversity. How else do ecosystems diversify? How can any animal survive otherwise?

In the case of our green turtles, consider this: if we fence off an area of seagrass and remove turtles for a couple of years, we wouldn't be surprised to find an increase in seagrass. Hey presto! We have solved our seagrass decline problem. Cull sea turtles and then we can look forward to more seagrass, more carbon capture, more fish and a brighter future for all humanity.

Except that's nonsense. Rather than improving our future through the power of increased seagrass, the removal of turtles would worsen this ecosystem that we depend on so heavily. Why? Without turtles we would see a decline in seagrass productivity, a lowering of overall nutrients, less structural diversity and a weakening of subsurface root structures. It does not need to grow back stronger and healthier as there is no predator for it to resist. The consequence of this is a lowering of resilience to storms in coastal areas, a huge decline in the ecosystem's life-carrying capacity and much-reduced diversity of animal life.

Turtles don't live in isolation. Herbivorous and predatory fish, sea urchins, dugongs, tiger sharks, dolphins, terns, swans and hundreds of other animals all make up a seagrass ecosystem. These naturally occurring hierarchies began millions of years ago and periodically settle into predictable patterns and a steady structure based on the collective knowledge and movements of all their components. All of this ensures the survival of these ecosystems and the individual components within it.

LESSON 2: Be like a turtle and create civilisations compatible with the animals that live around you.

Let's consider what would happen if we did remove turtles to allow seagrass to 'flourish'. The seagrass would grow back but it would be weaker, as we have acknowledged, and at some point we would need to reintroduce turtles to curb that seagrass growth otherwise it would never stop.

But if we reintroduce turtles into new seagrass habitat, they won't have the culture necessary to know how to forage efficiently. Without an existing steady structure and their peers to share existing knowledge about where best to find food then their efforts are, at best, random and ineffective[22]. The ecosystem would flounder because that turtle civilisation's connection to nature has been compromised. We have upset that balance.

In this respect Mr Titchmarsh is correct in saying that 'rewilding does not increase plant diversity'. By adding turtles back into this ecosystem their initial actions seem more like gardening. But perhaps we've got used to seeing animals behave this way because we mostly observe them in ecosystems we've degraded where their knowledge has been corrupted. They still need time to reconnect with nature like we do, to restore their culture and balance before diversity increases once more. Perhaps we think we're observing nature's brilliance in gardens when what we're really observing is another civilisation in pieces.

Gardens are not ecosystems. Ecosystems are collections of many animals other than humans that reorder broken pieces from each other's activities and create the circular economy needed to maintain long-term wellbeing. Gardens are disruptions to nature made by humans, created by destroying parts of an ecosystem to service a short-term need like food or wellbeing. Modern farms are scaled-up versions that have taken over whole ecosystems. That's not to say that either of these are unnatural; there is nothing inherently wrong with gardening or farming. All we need to understand is that without animals to counterbalance the elements of destruction, there is no mechanism to keep the supporting ecosystem alive and well.

When First Nations people first settled Australia it's believed they altered the landscape from heavily forested and megafauna-dominated to grassy woodland. They hunted, and the subsequent absence of heavy-footed animals contributed to a drying continent. Most megafauna went extinct about 25,000 years ago and the route back from this imbalance was tens of thousands years more. The time it takes to recreate a culture that is in harmony with the environment is in the order of thousands of years and that is how Europeans found Australia in 1788 – with an abundance of wildlife.

Mediterranean turtle civilisation didn't just spring out of nowhere, it was the extension of a knowledge legacy lasting for millions of years before. The human civilisation we point to – the post-industrial boom period of the mid to late 1900s, the one we desperately cling onto now – is only about 50 years old. Presumably, a turtle can't imagine taking all the seagrass for itself and creating a global trade in seagrass. Humans can, which is why our culture is more sensitive to corruption and our beliefs in our own constructive power easily fail us. Perhaps our inability to understand our place in the delicate balance of the ecosystems we live in is also why our civilisations are so prone to collapse.

Though it is a fact of life that we can't find food, water and shelter without disrupting ecosystems, there are ways to do this more considerately and consciously. As we increasingly try to garden all of nature, we become more removed from understanding our place in the ecosystems we exist within.

Alister Scott from the Global Rewilding Alliance tells of how he is 'trying to change the language of conservation from "restore" to "allow recover"'. Restore has the feeling that humans play the dominant and active role, but in fact the more humans interfere with nature, the slower we progress. Rewilding our world with animals allows nature to recover itself. If we rewild our world by reducing our interference with nature, conservation happens faster and reduces the demand on people, leading to more sustainable outcomes.

What if we stop making random choices? What if we don't listen to the media or take advice disconnected from our life-support needs? How much more should we be focused on the deforestation of our local watershed world politics? Or about housing estates replacing local food bowls? Or whether your cost of living, increasing insurance and mortgage debt is being driven by local ecosystem collapse?

Instead ask yourself what you know about your local wildlife, because that's what really matters to your future. All we need to do is to take our lesson in nature from animals like turtles to create civilisations that are compatible with the wildlife living around us.

Lessons from Our Connection to Creatures

The humble banana has twice as many genes as a human and that's not what the Human Genome Project expected. It turns out that our larger and more complex bodies need more disorder in their makeup. Why? Because we are more prone to collapse, so we have inbuilt redundancy to respond to outside influences by creating proteins in different formats to suit different scenarios.

This has led scientists to conclude that human bodies operate as an open system, meaning the environment around us helps determine how our body chemistry behaves. When you go out the people you meet, what you do and the climate around you define how your body reacts on a molecular level that day.

This is happening in the nuclei of every one of your cells right now. Your body is its own ecosystem and has just enough resilience to support life within the normal extremities of the outside world. Inside your cells there are also mitochondria. These originate from bacteria that formed a symbiosis with a long-distant ancestor and influence how you store and use energy. They are enormously affected by things like caffeine and while in nature it exists in moderation, most of you will have poured a distilled version into your body this morning.

In your gut and on your skin are about 40 trillion living organisms. Gut bacteria help store fat, make blood cells, replace damaged cell linings, influence the makeup of your body's immune cells and control your mental and physical health. They are influenced by what you eat; whether your food is loaded with pesticide, full of the wrong types of fat or a healthy and nutritious diet with plenty of fibre, your gut fauna help decide whether you're angry, stressed or calm. That also impacts whether you get heart disease and other ailments. 'Fifteen percent of the small molecules in our bloodstreams come from gut microbes, and those interact with every cell in our body,' says Kristin Ohlson in her book *Sweet in Tooth and Claw*[23].

There are also parasites that try to live on and in you such as tapeworms, ticks, fleas and so forth. Mostly we avoid these by living a clean and healthy lifestyle. Still the increasing density of our living spaces coupled with our wasteful existence has led to at least one-fifth of us becoming infected by the brain parasite toxoplasmosis – an organism that we tolerate and may even depend on. These parasites could help determine whether you are a successful businessperson, liberally minded or compassionate. But as we'll see, our levels of wellbeing can become unbalanced if we disconnect too far from nature and allow the parasite to take over.

Then there are the animals you live among that build the ecosystems that provide the food you eat, that fuel the gut bacteria and that energise the mitochondria to make proteins so you can live. A study of 16 mammal and two bird species from camera traps in Wisconsin has shown a far greater propensity for animals to congregate where human disturbance is high, such as in heavily urbanised or intensive agricultural areas[24]. This is a two-way relationship where balance and cohabitation are key to the survival of all within a particular ecosystem. The extent of that is evidenced in the fact that there are about 80 species of terrestrial vertebrate living within 1.6 km of my house. If you want to find out what lives in your neighbourhood like I did, I'd recommend signing up for iNaturalist and start exploring.

That outside influence extends even further to where your food, water and shelter come from. For me, that's about 300 species in the vicinity of the forest catchment that help deliver clean mountain spring water to my tap. Yes, here in Melbourne we drink water that comes from a mountain. In my humble opinion it's the best tasting tap water in the world!

All the animals that live in these huge energy networks influence us from the outside in. We are part of this network but don't realise what a small cog in this impressive machine we actually are. Animals cooperate with each other and us by sending signals in the hope we can all live in harmony. After all, if every animal was constantly fighting how would we all have the time left to do the things we need to survive?

Without climbing inside the mind of another animal we will never know exactly how it feels, thinks or communicates. Nevertheless, animals 'talk' to each other and share knowledge, find ways to fit around each other and maximise the output from ecosystems by reducing conflict. This doesn't mean they need to be able to speak, instead they communicate through direct and indirect signals.

When a whale hears a vessel engine approach it shouts the only way it knows – by using a thunderous clap. It may be the loudest natural noise in the ocean and lets them pinpoint each other's position. We don't need to know what they are saying. It's enough to know that a sentient creature has waved to us. We should already know what to do.

This happens in humans, too. Just think of the phrase 'actions speak louder than words'. Some of the ways we relate to and influence each other's behaviour are emotional or based on a body language we can all understand. We don't have to punch someone in the face when we're angry, nor do we have to verbally explain why there are tears falling from our eyes when we're sad for someone to understand the emotion we're feeling.

It's more than just communication. The way we feel and react is driven by factors more complex and numerous than we will ever know.

Humanity's existence is a complex interplay between all living organisms. These are powerful natural processes that determine our destiny and make us a slave to nature and indeed, our own nature. The lesson is to accept that our future is built around influences we can't control and that no amount of research is going to reveal all of these to us. Indeed, we hardly know anything about the most sinister and obvious relationships we have with our environment and we aren't even aware of many more.

Our next set of lessons from nature focuses on the close relationship we have with animals and how these outside influences can impact our lives and bodies, connecting us from the smallest strands of our DNA to the mountains that provide our drinking water.

Cats, Parasites and the
Wolf of Wall Street

For Mother's Day I bought my canine-loving mum a chance to walk dingoes at a conservation park near Melbourne. She was thrilled. Throughout the walk the dingoes behaved just as dogs will, investigating invisible odours until inevitably one rolled gleefully on its back and writhed in delight, it's woolly mug almost grinning as it spread itself in dung before proceeding to eat it.

At some unknown moment in time, perhaps even before wolves evolved, a dog-like animal did just this. There are several possible reasons why – maybe it helps with gut bacteria, or perhaps it conceals the dog's scent when hunting? Either way, that dog-like animal ingested cat faeces and set off a relationship between pets and people that is perhaps the weirdest of all animal-to-animal relations. They say three's a crowd; well, when the third party is a brain parasite it even crowds one's judgment.

One in five of us reading this book has toxoplasmosis of the brain. You are most likely to have caught it from a cat, and while it doesn't make you feel obviously ill it does make you behave differently to other humans. In fact, two papers published recently support the idea that this common brain parasite can even make those who are infected vote differently. It can also make you less cooperative and a bit more aggressive, which in wolves and humans makes us more likely to become the leader of the pack.

As well as manifesting as changes in levels of aggression, toxoplasmosis can also increase antisocial behaviour in humans. Ironically, nature has decided these qualities are also what makes a good businessperson. This is a wonderfully quirky example of our co-dependence on wildlife and how much we as individuals are controlled by those outside forces of nature.

Cat owners will already know that cats are capable of mental mind control but perhaps there's a reason why this view has made its way into common consciousness. To understand this weird relationship we need to look to wolves. In one study, scientists looked at 22 years of blood data involving over 200 wolves in Yellowstone National Park[25]. Wolves that spent time near cougars (which are cats, though admittedly rather large ones) were more likely to be infected by toxoplasmosis. This meant that:

- Both male and female infected wolves were more likely to leave their pack earlier than they normally do.
- Infected males were an amazing 46 times more likely to become pack leaders than uninfected males.

While this type of parasitic mind control is not unusual in the animal world, it is only quite recently that scientists have been seeing an uptick in toxoplasmosis in humans. As Imperial College London scientist Joanne Webster says in an article for *Discover Magazine*[26], 'We often see symptoms like altered activity levels, changes in risk behaviours, and decreased reaction times … but in some cases, they become more severe – like schizophrenia.'

Those changes in risk behaviours could certainly have an impact on a person's business acumen. It would seem likely that another wolf – this time the Wolf of Wall Street, Jordan Belfort – had a soft spot for cats, as even his latest ventures include a blockchain-based pet company. I'd bet he grew up around animals and probably, inadvertently, consumed some faeces when he was young. Though it should be noted that this isn't something we should be aspiring to – nor will it make you a millionaire – so please don't try it at home.

The association between toxoplasmosis and businesspeople and entrepreneurs is well known. A study in 2018[27] found that:

- Students who tested positive were 1.4 times more likely to major in business.
- Professionals attending entrepreneurship events who tested positive were 1.8 times more likely to have started their own business.
- Nations with higher infection had a lower fraction of respondents who cited 'fear of failure' as a factor inhibiting new business ventures.

Your chance of being infected with toxoplasmosis could be higher depending on where you live. In the US and UK that chance ranges from nine to 11 per cent. Those of you in Canada and Australia have a higher possibility of contracting the infection at 20 to 23 per cent. New

Zealand is higher still at 35 per cent, with Indonesia and France taking the leap right up to around 54 per cent. Cosa Rica tops the toxoplasmosis-contracting table at a whopping 76 per cent[28].

What isn't as well known as the businesspeople–toxoplasmosis connection is the effect that toxoplasmosis has on our political behaviour. But if we know that leaders are more likely to carry the parasite, what does that mean for the rest of us?

A 2021 study in *Evolutionary Psychology* looked at a sample of over 2,000 Czech men and women[29]. The findings showed that 'men and women infected with the disease demonstrate lower conscientiousness, generosity, and novelty-seeking'[30] and also had poorer health. It also showed a tendency for people to be less loyal, less liberal and more anti-authority. This change in behaviour is thought to be related to inflammation of the brain that occurs because of the infection.

For the most part these changes are quite benign and only at the extremes do we see the worst behaviours emerging. There's an amusing story in the book *Fifty Shades of Gray Matter* by Teresella Gondolo[31] about an Ecuadorian man who used to rescue stray cats. His parents named him Hitler (no, I'm not making it up) and he was hospitalised with uncontrollable arm movements that resembled a Nazi salute. He was identified as having hemiballism, which comes from the Greek meaning 'half-jumping' and describes a condition where someone makes involuntary movements. This condition had come about from developing a brain lesion after contracting acute toxoplasmosis. He since recovered.

The real Hitler hated cats but was madly fond of his dogs, which is another way for the parasite to reach the human mouth. Viable toxoplasmosis oocysts are found in dog fur and can be ingested after petting an animal that has rolled in cat faeces. Was Hitler infected? We will never know. Though it does make you wonder, doesn't it? Why is there always a cat in Number 10 Downing Street? James Bond's arch enemy Blofeld was famously depicted with a cat. Could he have been heavily infected with toxoplasmosis? Studies suggest so.

The idea of cooperation in nature is a theme in Kristin Ohlsen's book *Sweet in Tooth and Claw*, which explores the relationship between people and microbes, the majority of which are good for us. The simple fact is that cooperation is what enables populations to survive as the more aggressive and dominating portions of society die out fast.

If this is the case, then how do less cooperative behaviours persist and why? If toxoplasmosis makes people less cooperative, what role does it play in nature?

At the most basic level of our understanding of human society we know that conflict drives us forward. The least cooperative die out the quickest, but the more cooperative portion would stagnate without something to struggle against. Another way of saying this is the battle that socially woke (here I use the real-world definition of woke) people perform each day for a more just society could not succeed without something to fight for. Equally, the uncooperative people they are fighting rely entirely on this socially progressive, loyal, peaceful and cooperative tribe to survive.

Consider superheroes. Supervillains are the ones always pushing to change things, to go against the norm or what is considered socially acceptable. Superheroes battle against them, working to maintain order and peace on behalf of this wider population. Without this tension nothing would happen – we would all just bumble along in our own worlds and everything would stay the same. But the actions of those supervillains make us think about and question what we perceive to be the standard and from that comes progress.

Another way to think of it is like a chariot moving forward. It's driven by horses, forced into a gallop by a whip-wielding lunatic who, we all agree, is not the nicest person but without them the carriage grinds to a halt. The chariot delivers what we need to survive, but in the necessary rush to get there the wheels start falling off and the coach begins to disintegrate. The rest of us run behind, picking up the pieces and rebuilding the chariot. Some manage to climb on, overthrow the lunatic and take over, but then they become the new face of lunacy and so the cycle continues. The chariot never actually reaches anywhere. All we know is that if we race too fast it falls apart. If we slow the chariot down, keep it moving forward and make sure the drivers are as sane as possible it lasts longest.

It is a little absurd to think that a parasite of our brain might exist to create aggression and, in doing so, help our society to survive. But this is how and why toxoplasmosis lives in the great balance of the natural world. By existing in the back of our minds, so to speak, it forms part of the unconscious contribution made by cooperative and uncooperative humans to forming a balanced and enduring society.

Being infected with toxoplasmosis isn't necessarily going to make you into a despot, because the world is a kaleidoscope of different overlapping values and cultures. You may have been chosen by chance (or more likely by a cat) to be bit more aggressive than most, but that's not a bad thing. You still function as part of ecosystem processes you aren't even aware of.

> **LESSON 3**: Animals connect to us all the time, inside and out, shaping our destiny by influencing the way we think, act and behave.

Early evidence of toxoplasmosis came from the DNA of an Egyptian mummy. It was about that time that wild cats became domesticated because an increasingly dirty and wasteful human living environment created opportunity for rats to thrive. Cats eat rats and the rest is history. Cats have been part of our culture for tens of thousands of years. It's no accident that Ancient Egyptians thought cats had divine energy – that would be the toxoplasmosis talking – and 10,000 years would have been long enough for there to be an association made between cat owners and more successful businesspeople. Revering cats became a case of group natural selection where a heady mix of animals living together created ideal conditions for survival. In this case a cat, a human and their pet parasite.

Jump forward to the present day and Western Civilisation dominates the globe. Yet while we may argue about differences in political ideology, we are all the same in the eyes of nature. Toxoplasmosis is having a renaissance and it isn't too discerning about who it chooses. Its prevalence in South America, where three-quarters of the population might be infected, can easily lead to an overall angrier and less liberal society. Could this pesky parasite be behind the rise of right-wing extremism? While we can't lay the blame for this entirely at its door, it could be partly to blame.

In one of his first public appearances since being elected president of Argentina in December 2023, Javier Milei (who many Argentinians call 'the madman') shouted, 'I'm the king of a lost world! I'm the king and I will destroy you!' He also believes he's on a mission from God and can communicate with him through his dead pet dog ... who he has had cloned so he can still give him political advice. These are the kinds of situations that make you wonder just how often Milei's beloved pet enjoyed a good roll in the dung.

Cooperative or not, leader or follower we cannot escape the fact we are all completely connected to nature and often in ways that are somewhat unsavoury to think about. It's interesting and a little worrying to watch a growing mass of people angrily idolising the political views of those whose behaviour shows a blatant disregard for our place in the natural world.

Toxoplasmosis has always been a regulator of civilisations, and complex animals – whose role in life is to build and maintain habitable ecosystems – are the perfect vehicle. This is a parasite that infects the brain but doesn't kill us, an organism that has exploited the need for animals to work together for the benefit of the planet and forged its own survival as part of evolution. And here is another lesson: toxoplasmosis will play its role in resetting Earth's imbalance and putting humans back in their place.

It is high time we realised that people and politicians have never been in control. Instead, how we behave and our society's destiny are mere vehicles for ecosystems ruled by animals – some of them invisible to the human eye. Cats are only one of countless examples, but we notice this connection because the animals we live closest to are the ones that have the greatest influence on us. Did we domesticate cats? No. If anything a parasite domesticated us by making us love cats. Now is the time to acknowledge that nature is all powerful and these strange human–wildlife relationships endure because they have always been the most likely to help us survive.

Balzac, the Birds and the Bees

I was drafting this story sitting outside a café where I was consuming caffeine. In the neighbouring eucalypt, noisy miner birds were busy feeding on nectar from flowers that also contain caffeine. Meanwhile, the smaller thornbills were feeding on the insects that eat the nectar that contains the caffeine. Buzzing from my beverage, I began to wonder whether the consumption of this substance created a level of fastidiousness that has sped up adaption? Do we have natural caffeine to thank for the rate of our own evolution as a species?

Most animals – humans included – use caffeine to accelerate their capabilities. Everything that consumes it finds a burst of energy and gets on a metaphorical treadmill that (in a well-balanced ecosystem) rapidly diversifies vegetation. This in turn leads to flourishing ecosystems that deliver life support for animals. The circle of life is taken up a notch when caffeine is involved and it could easily be argued that this substance is a big part of the universal chemical architecture for all life on Earth.

Caffeine is the most ubiquitous drug on the planet, infecting every animal that encounters it. While plant caffeine is extremely poisonous it is also highly addictive. It will make you think you can't live without it, then subsequently kill you if you have too much. The thing is, plants wouldn't even contain caffeine if they weren't grazed, which immediately establishes that plants depend on attracting animals.

As with everything in the natural world, balance depends on pressure from the top down. It's not so much a reciprocal relationship between plants and animals as one where plants depend on animals for overall ecosystem stability. Humans happen to be animals. While some plants could theoretically exist on a planet without animals, a planet that has humans must be one where plants depend on animals.

Caffeine has no other use but to tempt animals to have a munch on plants. It's even stored in areas separate from the rest of the plant so it can be readily found and consumed. However, for the sake of balance it's also poisonous in large quantities. The pressure from animals eating it has forced plants to evolve a mechanism that regulates consumption so the ecosystem can be as productive and diverse as possible.

Plants can't exist alone; they need animals to create ecosystems by diversifying plants to provide functional life support for other animals. Put simply, plant diversity is a by-product of animals building their own living environment. The existence of caffeine is necessary for plants to survive, as without animals grazing upon them they would descend into a monoculture and many caffeine-containing plant species would go extinct.

Humans, just like any other animal, are hooked on this drug. While science is undecided on the benefits, 90 per cent of us regularly consume some form of caffeine. Health experts suggest you shouldn't consume more than about 200 milligrams at a time, which is a little under what you get in a Starbucks Grande Caffè Americano. A study of 350,000 people found moderate and habitual coffee drinking helped avoid heart disease[32]. It was recently added that regular coffee drinking limits reoccurrence of bowel cancer, while other research says it poses a risk to unborn children. There are so many studies.

The fact is we should be less concerned about individual consumption and more about the effects that imbibing this widespread and omnipotent drug might have on our society. Given that caffeine is a vital component in the creation of ecosystems and a driver for how populations of animals can coexist, that means that although individual consumption is a health issue the population-level consequences are far more ominous. Forget the post-coffee jitters, caffeine consumption has another nasty surprise for us.

Caffeine is a psychoactive drug that binds with chemicals in the mitochondria that live inside every one of our body's cells. These little fellas are responsible for turning food into the energy that makes our bodies work. Caffeine assists that process and wakes you up by increasing your brain's energy levels. With caffeine you're more likely to be better at simple tasks that require constant and repetitive work, such as entering data into a spreadsheet or laying bricks to build a house.

Mitochondria have a specific genetic code that is only passed down from your mother's side, but if we look a little further back we find they originate from when a microorganism formed an interaction with a bacteria hundreds of millions of years ago. Every single animal on Earth evolved from that original pairing, meaning caffeine can affect all animals in the same way – it instantly permeates almost all the cells in our body all day, every day after it's consumed.

In *This is Your Mind on Plants*, Michael Pollan (2021) describes caffeine as one of only three substances that have had the most profound impacts on human culture and ideology. From wars to philosophy and everyday work, our whole society runs on our addiction to caffeine.

The French novelist Balzac also knew this. In 1838 he wrote this disturbingly contrite essay on the effects of coffee.

In the main, coffee taken on an empty stomach puts you in a sort of nervous liveliness resembling anger. One's words rise and gestures express a morbid impatience to resolve a multitude of ideas that float around. We become brazen and angry over nothing. We become The Poet, he who irritates the humble grocer. Though we fancy the grocer is actually enjoying our company and sharing in the lucidity. A man of intelligence must, therefore, be careful not to show himself or let himself be approached. I soon lost, by coincidence and without effort, the delight I thought I had obtained. Friends, with whom I travelled in the countryside, found me surly and argumentative, discussing matters in bad faith. The next day, I recognized my faults, and we looked for the cause. My friends were scholars of the first order, and we soon found it: the coffee wanted its prey.[33]

Le café voulait une proie, says Balzac. The coffee wanted its prey. Here in broad daylight is a whimsical but worrying portent of the power of caffeine (indeed any drug) to both give and to take away aspects of our society.

As many people get more tired, they drink more coffee to accommodate demands on them but this only gives the impression of efficiency. What caffeine really does is increase anxiety and reduce creativity, becoming a self-defeating philosophy where our more creative elements such as storytelling and other aspects critical to any surviving culture become compromised.

As Pollan says,

> Caffeine improves our focus and ability to concentrate, which surely enhances linear and abstract thinking, but creativity works very differently. It may depend on the loss of a certain kind of focus, and the freedom to let the mind off the leash of linear thought[34].

Perhaps it's not the best companion to writing a book after all …

In a well-balanced ecosystem there is such diversity that it is impossible for caffeine to dominate. Over many tens of thousands of years ecosystems forced animals into a life where they consume just enough of it to maintain that life-giving equilibrium. Disruption, however, can tip this balance and this is what humans have done through mass-scale farming of coffee beans and concentrating caffeine through mechanical devices. Even though we tend to drink less coffee than we did in the middle of last century, the amount of caffeine being consumed globally (soft drinks included) has doubled since 1990.

As with blood sports and trophy hunting, our commoditisation of this incredibly powerful aspect of nature is not connected to any positive ecosystem outcome. The real risk with caffeine is that it isn't that poisonous, which means though it's not enough of a danger to each of us individually, overall our general consumption of it is a clear threat to civilisation.

LESSON 4: Working too hard isn't the same as being productive.

There are thousands of papers looking at the caffeine as a pollutant or medicine, analysing chemical pathways, effects on sleep, locomotion or behaviour. For me, there is only one study of note and that is the one that connects caffeine to wildlife colony collapse. In nature and away from the influence of those pesky scientists, it's highly unlikely that any wildlife population would be able to overconsume caffeine, which is why the work of bee ecologist Margaret Couvillon is so interesting and unique.

Couvillon led a study that modelled the effect of what would happen if plants tricked bees into using caffeine more frequently. Plants are already like little baristas, enticing bees to drink our equivalent of 200 cups of coffee a day. Couvillon's team investigated what would happen if bees were artificially enticed to come back time and time again to the same flowers. The results were intriguing, showing that too much caffeine leads to the bees overestimating the quality of the flowers. As a result, their foraging behaviour becomes erratic, leading to a decline in honey production and ultimately colony collapse.

While there is an important lesson to be learned here, there's no need to throw out that bag of coffee beans just yet. Chaos and collapse don't happen fast – it's a slow demise through gradual disintegration of a population rather than the poisoning of individuals.

Five grams or more of caffeine can lead to an overdose and death, which is equivalent to about 50 average cups so it's very unlikely that we're going to be picked off one by one. About five cups a day is enough to cause increased anxiety, which can lead to stress. Two to three cups a day is considered safe, but how much on average is okay for a society to continue to function sustainably? No-one knows.

What we do know is that working faster and longer doesn't mean being more productive. In fact, it could mean the opposite as this would likely lead to burnout making to harder for us to do anything effectively. Both Balzac and the bees knew this, and it should give us pause for reflection. The answer doesn't lie in banning coffee but in changing the structure of society outside our bodies to enable less coffee drinking.

If we cared more about our employees they wouldn't feel the need to consume as much coffee, which wouldn't lead to stressed, anxious and burned-out staff. Instead, we could have just enough to make us more efficient and dynamic, striking a balance that would result in us being healthier, happier and more productive as a society. It's about fostering the right belief system to allow society to thrive as nature intended, rather than leaning too heavily on a ubiquitous and socially accepted drug.

But what about the bees, you ask? And what could their colony collapse teach us about our own survival as a species? If a bee colony verges on collapse, a gap in the ecosystem is filled by another subpopulation that becomes dominant by behaving better. In other words, this new population has a bit more self-restraint than the heavily caffeine-addicted bees. Everything ends up okay again. Balance is restored. This is the natural state of things and in the short life cycle of bee civilisation, it might happen very regularly.

The fast turnover and adaptability of animals in ecosystems is one of the reasons they provide us with so much resilience. Insects breed fast, make decisions fast and generally occur in abundance, which makes them powerful components in ecosystem regulation.

While our human civilisations operate on far greater timescales, one might argue that what makes bees susceptible to colony collapse might also make our society more vulnerable to the influences of something as addictive as caffeine. Even though we don't have specific studies to prove this, we do find time and time again that we are made similar to all other animals – including bees.

When a songbird raises chicks, it's less likely to abandon them after fledging as their investment is already too great. Even if it means raising chicks that will almost certainly die, they will persist. Humans are no different. We've also invested greatly to get this far and will persist in living the life we have established – we will even vehemently defend coffee drinking to the death.

These natural vulnerabilities are what make us slaves to the power of nature that forces animals like us into submission. It also stands for our relationship with caffeine. When we drink too much caffeine and our society makes chaotic choices to stray beyond the limit imposed by outside influences, nature is there to hold us accountable to bring ecosystems back into balance.

Coffee is neither good or bad but in revering it and depending on it we have become slaves to a power greater than us and have allowed our society to drift out of balance. It's an outcome that can only be tempered with an economic contraction where the wholesale addiction to caffeine will naturally wane over time. Caffeine consumption has almost certainly led to the rapid rise of technology and human achievement and our generation has benefited enormously, in so many ways. It will just as inevitably lead to a reset of our society as return to order, restoring ecosystem balance and wildlife recovery as we move further away from consumption of caffeine.

The bees I watch as I write this aren't thinking about any of that. They're searching for food, not thinking about getting a caffeine hit. They are doing what comes naturally and drinking just the right amount to start rebuilding our world. If all the bees in the world thought too hard, they would forget to follow their instincts and their procrastination would lead to the breakdown of ecosystems. We, too, are in danger of forgetting to do this by allowing our need for a daily cup of coffee to overrule human nature.

Abundant wildlife behaving normally is essential to the long-term success of any species, humanity included. Our living environment will become more habitable again once this interaction between humanity and caffeine is rebalanced to provide a more sustainable existence.

Sympathetic Stingrays and Adventitious Apes

From above the water, we could see pelicans, cormorants, terns and gulls feeding on a densely packed shoal of small bait fish. As we slipped beneath the waves, streams of foot-long predatory salmon were piling past. I could feel them bouncing off my legs before they smashed into the pulsating disco of small fish in front of me. Then, suddenly, an enormous ray appeared.

Smooth rays – colloquially known as 'stingrays' – are animals that Australians use to strike fear in visitors to the country, especially after legendary conservationist Steve Irwin was killed by a related species in 2006. In truth, Irwin's unfortunate death was a rare incident, and stingrays are gentle giants. The one in front of me was two metres from nose to tail and a metre and a half across, its muscular tail sporting an unsheathed barb the size of a large carving knife.

After the ray passed I stood up on bare rock and it turned back towards me. Ever so slowly this huge sea pancake rippled forward before slowing and gently resting its head on my feet. For a few moments we contemplated each other eye to eye – this old, inquisitive and intelligent animal looking at me with curiosity and knowing. There I stayed, taking in that moment where a majestic animal such as this was greeting me like a dog might do in a local park. Here was language that transcended speech – an acknowledgement of mutual respect like a handshake or doffing of the cap.

We parted company after a moment, and the ray proceeded to feed while I took photos. As if to reinforce trust, it repeated our encounter about 10 minutes later. By that time my mind had moved on from simply appreciating the present moment and had begun contemplating the deeper meaning behind this close encounter.

What knowledge did this stingray have about the ocean that I would never know? How much impact has it had on the health of our bay since it was born and swum away independently, likely decades before? What did this show of mutual curiosity and respect mean for my understanding of the human–animal relationship on this planet? For one thing, it renewed my desire to protect the local ecosystems and these age-old rays.

Animals aren't averse to seeking human company. From the blackbirds that follow you digging in the garden to the coyotes that live in the US suburbs, stories of human–wildlife interaction go back centuries. There is the Biblical story of Saint Jerome removing a splinter from a lion's paw while living as a hermit in the desert. Don't believe it? What about Cristina Zenato in the Bahamas? She's put her arm into the mouths of over 300 sharks to remove fishing hooks in recent years. 'Once we relax, they relax,' she says. 'I observe their behaviour and energy and they can feel the range of communication without one word exchanged between us.' Then there's the story of Pocho, the wild crocodile nursed back to health by local fisherman Gilberto Shedden. After that encounter, the crocodile sought his company and they lived together for 20 years until Pocho died of natural causes in 2011.

Anthropologist and primatologist Biruté Galdikas began studying orangutans in the wilds of Borneo in the 1970s. Her relationship with these primates was probably the first time a human and orangutan had ever communicated that way. A large male named 'TP' could have easily killed her, but Biruté protected herself from this threat by keeping a respectful distance. Over time trust was established and TP allowed her to approach him and get closer than perhaps any orangutan had ever voluntarily allowed a human – but not too close. 'Clearly, TP had become habituated, but so, I realised, had I,' she says in her book *Reflections of Eden*[35]. 'The process was reciprocal. Gradually, TP and I worked out an unspoken agreement.'

To confirm the way any animal behaves takes at least two generations – for female orangutans that's 16 years of study that would be required. It took Biruté over 20 years to even begin to understand these creatures, but before this it was a widely held belief that orangutans couldn't be studied. Many men before her had tried and failed to get more than a glimpse of any wild orangutans, let alone a male.

That Biruté was possibly the first non-orangutan species to have ever communed with a male orangutan since they evolved millions of years ago should be considered a pivotal moment in the whole history of humanity. For me, this is more historically significant than walking on the moon. Biruté was on the cusp of a change in the relationship between humans and Planet Earth that signified the interdependence of species. That it happened with one of our closest living relatives and bridged the gap in communication between apes and humans represents a new version of that 'giant leap for humankind'. Biruté had started something special.

In revealing themselves to us, orangutans have had to adapt their behaviour to consider the new idea that some humans are not a threat. More recently, orangutans have been observed approaching people on the margins of palm oil plantations to ask for food. Ignoring the awful legacy of palm oil plantations for a moment, clearly at that location and time they didn't feel threatened. It is perfectly natural for wild animals to feed cooperatively. So when two highly intelligent primates come together, seeking food from each other seems quite natural. Perhaps more of this new relationship is what we both need.

The truth is we need orangutans as much as they need us. Even the Dayak tribes of Borneo are having to change their way of life to help protect huge swathes of forest. Their ancient animistic religion of Kaharingan was based on a deep spiritual connection with the forest; they knew why this ecosystem was important but didn't know how orangutans made it fruitful. Without this knowledge they hunted orangutans for food, leading to the orangutans going away to find their own space. Both species survived, but now things have changed. The forests are being cut down, meaning there is nowhere for the orangutans to hide anymore. Now, some Dayak tribes are realising they need orangutans to rebuild forests and that these animals are a vital part of preserving this delicate ecosystem on which their lives depend. These tribes are compromising their own traditional hunting behaviour to live side by side with orangutans in a different way – by becoming involved in their protection in a bid to preserve nature, their culture and combat the increasing demands placed on their forests by modern companies.

This changing nature of the relationship between humans and animals is not specific to the Dayak tribes. Cristina Zenato, who has been involved in shark conservation for 30 years, says, 'One of the most noticeable changes is in people still afraid of sharks who are now advocating for their conservation because they understand their important role.'

It proves the uncanny awareness of animals when it comes to our shared natural world and their understanding of the need to communicate and adapt our behaviour to survive together. The relationship between humans and any kind of wildlife is not fixed; even after millions of years and within a very short space of time, any two species can quickly learn to respect new boundaries of coexistence.

The wisdom offered to us by anyone so bold as to devote themselves to observing wildlife like Biruté reconnects us to an important part of our own history, culture and humanity.

> **LESSON 5**: Given the chance, wildlife will choose to engage with us and lead the way towards restoring ecosystems.

The relationship between humans and wildlife is like a well-performed tango. Both dancers need to move together, countering the other step for step to produce movement that is harmonious and balanced. But there is always a leader, one dancer that guides the movements of the other to ensure they stay in perfect synchronicity. This is the lesson we must learn from Biruté and the orangutans: we must let wildlife lead.

If we let them, animals will seek us out in a bid to restore anything out of balance in nature and reconnect with humankind. This is why human inaction may be the best form of action if we are to survive the next 100 years.

In the very recent past, scientific publications decided that conferring human characteristics onto animals was ungodly because to anthropomorphise animals suggests they might share culture, feelings, emotions and empathy. In turn, this would diminish our perceived superiority over any other species. The most learned behavioural ecologists studying animals instantly dispel this notion, but it remains the dominant opinion among many of the academic elite. Keggie Carew in the wonderful book *Beastly* quotes Charles Foster who says about animals that we are 'a rolling conversation with the land from which [we] come' and that 'all around us are the billions of individual selves acting out their dramas in their individual worlds, connecting to bigger and bigger worlds, to the higgledy-piggledy live jigsaw'[15].

Carew reminds us how this assumed superiority and inability to let wildlife lead has denied us learning. We have lost the chance of exploring possibilities and beliefs that fall outside of a current way of thinking, all because the dominant scientific narrative tells us that we are better than animals and therefore they can't offer us anything new or useful when it comes to understanding our place in the complex natural world.

Again, this is where it is pertinent once again to look away from modern scientific 'knowledge' and instead turn to indigenous wisdom. The intangible beliefs and values of indigenous cultures along with thousands of years of 'rolling conversation' that endure through paintings, dance, music and storytelling are a critical component in cutting edge conservation and community economic development. For some it remains a leap too far, but attitudes are changing.

In many indigenous cultures animals are considered spirits of ancestors. This derives from the type of empathy Carew describes in *Beastly*, and as Biruté says, 'When we're in the presence of animals, we have a deeper understanding of spirituality, of where we come from and where we're going.' Wildlife permeates our past, present and future selves because our very existence depends on living with and being connected to these creatures. After all, it takes two to tango.

The most successful societies observed wildlife closely and lived with it. Some engendered respect, care and reverence for wildlife so that everyday decision making – even on killing and eating animals – was done as though one was protecting one's own family. However, hunting is an instinct for humans and local fauna is often wiped out after settlement. It can take many thousands of years to reform the balance but this depends as much on the will of the wildlife as it does on the will of humans.

Animals choose whether to avoid or seek human company based on how threatened they feel and it is this action that will dictate the future of that delicate relationship between humans and wildlife. Allowing wildlife to lead can open the door to deeply spiritual interactions with humans. There are countless stories like the one I recounted at the start of this chapter where I shared a significant moment with a stingray.

One of these is from environmentalist Ray Lewis, OAM, who told me of a time he was snorkelling in the shallows and a fiddler ray was lying in the seagrass. She was obviously pregnant. He'd been visiting the site for decades and although he knew he shouldn't, instinct took over and he reached out a hand to gently caress her back where her young were in the womb. Realising his error of judgment he pulled away, but they ray did not. Instead, she swam up to his chest, placed her head between his arms and lay on his belly.

'It was a moment that brought me to tears,' he says.

To this day he can't explain what happened but for a moment two intelligent animals trusted each other and shared a connection that went deeper than words.

In Eastern Indonesia, my friend Edi Frommenwiler (who owns the ship Pindito) was diving with a group when they found a turtle tangled in fishing line. The divers freed the stricken animal and as soon as it was released it swam quickly away. But then it paused, turned around and circled the group a couple of times before disappearing for good. A signal of gratitude, no doubt.

Some animals even seek out humans to assist when they are struggling. During bushfires in eastern Australia, koalas climbed into buckets of water left on people's verandas. Others, like primates, can form lifelong bonds or friendships with primatologists and recognise them after years of absence.

Animal brains float alongside each other and our interdependence means that while we may not share a language, we do have to be able to communicate quite well to avoid a collision of cultures. Who chooses who? Given that we're not in the position to force any wild animal to be our companion, it surely must be the animal that chooses us.

When an animal does choose us – as my stingray friend did with me – and we let them lead, this is the point where our drive to preserve and protect the natural world comes bubbling to the surface. In the ABC TV film *Platypus Guardians*, Tasmanian Peter Walsh recounts a moment when a female platypus came to him. It chose a moment in which to confide in his presence, climbing onto the riverbank and preening itself while he sat alongside.

It was as if she was trying to say something. The more I saw her, the more she would zoom up to me. She would always come up and raise her bill out of the air. It felt like a doorway had opened, a magical portal into the world of the platypus.

This event wasn't a one off – it happened time and time again.

Such was the impact this moment had on his life that Peter began a quest to restore the urban river his platypus friend lives in. Since then, the whole community has come out in support of the campaign. Peter, an unassuming and quiet man, inadvertently created a movement that's resulted in the large-scale rehabilitation of a platypus river corridor. He has become 'The Platypus Man'.

All of this was developed through Peter allowing wildlife to lead, to come to him and forge a connection that his own actions would never have been able to create. The relationship started by the platypus inspired Peter to work to preserve that habitat. Was it a conscious move on the part of the platypus? Who knows. But the result was a greater consciousness and conservation of the habitat this animal needed to survive. Imagine how much progress we could make in conservation efforts if more people remained passive and let wildlife lead the way.

Throughout society there are tales of people who have been chosen by animals to represent them. Many indigenous cultures formally recognise these as totems. People were often said to have been chosen by the animals after experiencing a connection with them and would speak on their behalf, which makes sense. If the animals your community eat are also critical for ecosystem balance, you need members of your society to moderate that consumption by speaking on behalf of them. Not only does this engender respect and convey a sense of personhood on the animals, it ultimately delivers healthy wildlife diversity and stable ecosystems with abundant food and water. Throughout time, these moments created a stable kinship between different species and this respect led to a form of cooperation that is the basis for the richest and most prosperous ecosystems.

Our society is peppered with people who have been chosen by all manner of animals to act as custodians of their rights. There are people all over the world who develop an attachment to a species of animal at a very early age and go on to be advocates for their conservation. Many more of us can trace our love of wildlife back to a single close encounter when we were young. These moments can keep happening throughout your life if you remain open to letting wildlife lead. It's strongly embedded in our DNA to foster relationships with other animals, care for and cooperate with them.

While our culture is intact, our society has failed us. Our natural instincts, the ones that have allowed us to survive for so long, aren't reflected in modern laws. However, as we have seen, that is starting to change with rights for nature being increasingly built into laws all over the world.

Allowing animals autonomy to make their own decisions and to live alongside us marks the difference between a society that will survive and one that will fizzle and die. Their right to live among our society is their choice – it should not be something we force them into or remove from them.

Not every conservation effort we make should be ignited by the spark from a significant moment with one particular creature. Instead we must fight for every animal on the planet, as we have no idea which ones are the most important for our future. Rather than giving in to those who seek to destroy nature, we should find ways to reward them with a greater knowledge and opportunity to be inspired to conserve it.

Our relationship with animals is not a one-way street. Whether it be parasites or pests, dominant orangutans or cuddly rays, all these relationships are part of the world's beating heart, body and lungs and we must dance with them – letting them lead – if we are to have any hope of preserving humanity.

The Sparrows of Kabul

All last year we were living in fear of that which we thought soon might be
But those little brown birds thought this was
absurd and flittered around so free
For the sparrows of Kabul would not be too
troubled if Taliban came into power
They'd eat their bread and shit on their heads
much as they'd done here on ours.
'THE SPARROWS OF KABUL' BY FRED SMITH[36]

In 2021, Australian diplomat, author and songwriter Fred Smith was standing on the embassy roof watching the sparrows of Kabul. They were picking crumbs from among the razor-wire topping of the high walls that separated Fred from the Taliban-controlled territory beyond. Watching these birds had become a daily ritual for Fred as he helped negotiate the mass evacuation of visa and passport holders after the US handed control back to the Taliban.

> I would pop up on the Embassy roof at about midday each day to catch some sun and exercise. It was a relatively mild winter in the city which portended ill as reduced snowmelt meant another drought was coming. By mid-March 2021, the buds were starting to pop out on the trees and the sparrows were flitting around gaily. I'd been reading the poetry of Mary Oliver whose reverence for nature is balanced with a clear-eyed view of its cruelty and complete indifference to human suffering. There's something oddly comforting in that, and I wrote the line 'the sparrows of Kabul don't give a shit' in my notebook. [36]

One evening in May 2023 I attended one of Fred's concerts that featured his poems, songs and stories about the missions to Afghanistan. I found myself sitting with dozens of Afghanis who had worked for the Australian

Government for years. For Australia, this had not been a nation-building mission – as it was misrepresented by both media and politicians – but instead was an undertaking to support the newly elected democratic government of Afghanistan. Afghanis and Aussies did this alongside each other, building a relationship that transcended any thought of violence.

It's that respect and collaboration that is key to our survival. In evolution, either biologically or culturally, populations will always collapse if they exert only a single trait or dogma or try to force a specific outcome that fails to heed the balance of nature. Violence is not the natural order of things, and most people shy away from a fight. Therefore it is in the minority ways of doing things – if proffered peacefully – where the greatest progress is always made.

Even though violent or forceful traits die out fast in nature, paradoxically they will be ever-present too. In times past, tribes would have resolved conflict at their own borders (often defined by mountain ranges and rivers), and over thousands of years cultural traditions were built that reduced conflict and created stability. For individual villages surviving as they always have there would be too much to lose in changing the status quo. For the same reason retribution and revenge is not that common among animals because the stakes are too high. Only those that are overfed will try to upset the natural order of things. Perhaps this is where the expression 'fed up' comes from?

Nonetheless, behind the evolution of all species are the aggressive forces that drag everything forward in time. This conflict pulls elements of the habitable world apart, leaving a variety of animals (people included) to pick up the pieces and rebuild as it moves along. Some forceful – even violent – conduct will always happen to ensure evolution, but it is the collaborative and innovative harmony concealed behind it that makes our world truly magnificent and meaningful.

The unfortunate fact is that aggression is visible yet, like so much in nature that is positive and progressive, the sparrows of war that get on with rebuilding are inconspicuous. This is how our media becomes consumed by fear, violence and conflict, often forgetting about the much larger peaceful and respectful collaboration happening underneath all that. The most important parts of the ecosystem – the ones that nurture, maintain and protect – hide in plain sight.

Given the media's bias towards bad news, unless we tell stories about these positive things then we become prone to making poor decisions and becoming too conflicted. Some conflict is inevitable but if balance is maintained it's all just part of evolution. The cruelty and indifference to war that Fred observed among the sparrows is reflected in our own behaviour and is only occasionally out of balance.

The sparrows flitting among razor wire fences share more in common with us than we may think because they have long been a conspicuous part of our culture. They are considered the carriers of souls and in different parts of the world symbolise life or death. House sparrows live almost exclusively around humans, behave like us and share our fate. When we make our own lives more difficult with conflict or violence, these birds also become angrier and eventually disappear.

For example, scientists have shown that even increased urban noise breaks down the territorial patterns of birds leading to higher levels of aggression. If patterns of sustenance such as the stable, diverse and democratic human society to which the sparrows of Kabul were connected are interrupted, then these birds will begin to fight more for scraps.

Sparrows are part of the ecosystem just like humans and all other wildlife. War doesn't alter that, but it does upset the delicate balance needed to maintain that habitat. War changes our relationships with common birds and animals. Sparrows clean up our spilt grain and seeds, balancing the populations of rats and mice and reducing the risk of disease. They provide extra food for those cats that infect us with parasites, which in turn diversifies our nature through making some of us more extreme and aggressive and making our species more resilient, adaptable and peaceful. If we alter the balance of the ecosystem that sustains the lives of the sparrows through war or other conflict, we alter our own, too.

There is a good reason why Fred found the sparrows of interest in the context of living through one of the darkest days in recent Western history. Understanding birds is a survival instinct thousands of generations old that is hard coded into our DNA and our psyche. When ecosystems are in balance, wildlife forms invisible, interlocking and intensely nested territories of mammals (including people), birds, insects and even bacteria uniformly covering landscapes. They recycle surplus energy back into life-giving ecosystem structures and restore a sustainable level of harmony and diversity. Remove or change just one of these myriad factors and the delicate ecosystem loses its balance – as do we.

As we will discover in 'Lessons from the Air', ecosystem instability accounts for a third of the impact of climate change. Climate change, the Australian Defence Force states, will increase Australia's challenges for peacekeeping, peace enforcement, and conflict[37]. As the world becomes a more difficult place to live, so tensions will start running even higher. Appreciating and restoring nature would be the perfect counterbalance, nurturing us through this period of change.

Climate change comes from a breakdown of the barriers between diverse cultures (tribes) or populations of animals. It leads to a disintegration of the structures that stopped animals from being aggressive to each other. It makes animals less efficient at managing the 'energy' released from plants and this ends up in our oceans and atmosphere. A fair climate and a habitable planet aren't just about temperature – they are also a way of life.

The climate for peace was changed in February 2020 following Donald Trump's disastrous 'peace deal' with the Taliban. This put a swift end to the 20-year investment by 53 other nations, cut the Afghan Government from negotiations and released 5,000 violent Taliban prisoners back into the streets of the regional capitals. Australia was forced to intervene after nations more powerful and forcefully motivated (the Taliban and the US) ripped apart two decades of rebuild on top of thousands of years of tribal diversification. The ecosystem lost its balance.

LESSON 6: Rebuild, collaborate and avoid conflict like the sparrows.

While the relationship between sparrows and people might seem tenuous, it is not. We are joined with them in our tendency towards collaboration and rebuilding. Despite occasional conflict, people living harmoniously together and alongside other animals has always been a main part of the symphony of life.

There is a lot we can learn by watching sparrows. Simple truths that resurface from the depth of our soul that aren't easy to explain but are there in the natural world. For example, a way of being that wellness practitioners might sum up as 'mindfulness' that the sparrow embodies and seems to radiate from its spirited behaviour.

Our contemplation of nature has always been among the most cathartic of human pastimes, but it's getting lost as our lives become more complicated, uncertain and disconnected from the world. By remembering to take time to stop, consider and learn from the wildlife around us we gain some valuable lessons in how to ensure our future survival.

Art and music are just two ways to bring this contemplation of nature back into focus. *The Sparrows of Kabul* is about those harrowing few days when Australian troops joined the Afghans and the sparrows to pick up the pieces of a war-torn nation. Through complicity in wars we risk, (like the sparrows, perhaps) becoming the carriers of souls – the souls of millions of Afghans who didn't have the money, status or opportunity to escape.

These people now face a climate less certain than they did before, one where the dominant rhetoric is weighted too much towards violence and control and not enough towards the compassion and cooperation needed to rebuild their nation. It's this compassion that Fred's book honours, the commitment of Australians and Afghanis – 'the sparrows' – who worked relentlessly with each other through those final days. The events that led up to the awful situation in August 2021 were outside their control.

Common animals that live around us every day speak to a simple truth greater than our own politics. They reveal how violence is not the natural order of things, even if we feel it surrounds our society. They teach us about nature's resilience and nurturing capability; contemplating nature and animals can reveal just how much positive force surrounds us, which helps us to survive even the most difficult of times.

Many years ago, about a year before the attack on the Twin Towers, I made a friend while travelling on the Pakistan–Afghan border west of Peshawar. It turned out that he worked for the Australian security services, though at the time, he was on holiday.

After Putin invaded Ukraine, my friend pointed out to me that all the world's most technically advanced militaries could not keep control of Afghanistan, defended by guys with homemade guns shored up in caves – they, too, were behaving like 'sparrows' of war, if you like. Human nature is to create cultures that are as steely as the rocks that characterise their landscape, and overthrowing these cultures is like turning over a mountain. The same stands for wildlife.

Humans can't even overthrow the humble sparrow because they are vitally important for our survival. In 1958 Chairman Mao ordered the killing of millions of them, which contributed to one of the largest starvation events on Earth. In the years following, China was plagued by an unprecedented locust infestation that collapsed grain production. Years later, the sparrows are alive and well.

The steadfastness of all animal societies is built on their connection to landscapes that flows both ways – it is a foundation for life more powerful than we can imagine. Where politics and wars are ephemeral, culture is permanent because it's ingrained in our society. But humans are benign actors in the wider natural world – we are not as significant as we like to think we are. Our wars achieve nothing but to upset the balance of the ecosystem we live in.

The message in this for all of us is that there is no end to conflict in nature because it is part of the natural order of things. But this doesn't mean we need to create or instigate conflict; instead we should endeavour to be more sparrow-like and rebuild, collaborate and learn from our mistakes if we are to continue to survive. Like a sparrow picking up the pieces, you must do as much as possible for the people around you.

Most people in the world are busy getting on with the daily hand-to-mouth rituals of survival. But while we are occupied with fixing up the mess we make for each other and ourselves, what we mustn't do is lose sight of our collective responsibility to avoid war at all costs. Why? Because there is just enough natural chaos in the world to maintain the balance we need to ensure our survival on this planet. Any more of that and we risk destroying everything we have worked to create and maintain.

As Fred says, we

> need to appreciate good government, boring and expensive though it is, since, in a chaotic world, it's what saves our country from descending into tragedy like Afghanistan.

Sparrows could not imagine interfering in the lives of others or else they wouldn't be sparrows anymore. They ignore conflict and instead choose to collaborate and build together. We, too, must be more like the sparrows and shun conflict to maintain our place in the order of nature.

Lessons from
the Ocean

We know more about the Moon's surface than we do of our own deep sea.

Being landlubbers we tend to underestimate what happens below the skin of the ocean, a glaring error in judgement that could impact our ability to survive the next 100 years. Instead of disregarding the civilisations below the surface we need to start paying more attention to our unique marine creatures, especially those that live around our own coasts.

Jenna Rumney – daughter of the late John Rumney, who the *Cairns Post* called the 'Attenborough' of the Great Barrier Reef – described the UNESCO World Heritage site as the 'city under the sea' during a snorkelling trip to the Low Isles near Port Douglas in north-east Australia.

Recently I discovered a video by *Neighbours of Fish Farming* in Tasmania presented by Miriam Margolyes[38]. During the documentary, Margolyes describes how 'a farmed Tasmanian Atlantic salmon ends up with more chicken fat in its body than an actual chicken'. That lovely pink salmon meat wrapped up and sold with pictures of crystal lakes and rivers on the packaging comes from fish fed on the feathers, fat and macerated beaks, claws and guts of chickens.

This might seem a little gross and slightly mad, but the bigger issue is that entire ecosystems are dying because of this practice. Creatures such as the endangered Maugean skate are now in the spotlight as their risk of extinction accelerates, and blame for this diminishing population is being laid at the feet of the salmon industry.

Salmon farming causes sharp drops in the oxygen levels of bays and estuaries that leads to the collapse of Maugean skate numbers. Also, to stay economically viable the farms use underwater explosives, shotguns and beanbag shells to deter seals, killing many of these vital hunters in the process.

All of this is radically altering the profile of coastal food webs. It's hard to imagine a future for coastal communities whose bays have no oxygen left and predators such as seals are reduced in number. The fact is that this will cause an imbalance in these ecosystems that will result in untold economic loss.

The onus is placed on scientists to prove these economic losses before they happen, which is obviously impossible, and once a tipping point is reached there is no reversing the impact. The implications of this ecological point-of-no-return for companies is that by using their rights to overcome social license, they will ultimately fail their shareholders as those economic losses begin to happen.

Corporations have more rights to exist than nature. In fact, they have more rights to exist than you and me (we will explore this more in 'The Surprising Elephant Economy'). Companies found guilty of killing people are routinely let free to continue operating; one of the most famous examples was the toxic oil leak at Union Carbide's plant in Bhopal, India that was thought to have killed between 3,000 and 16,000 people. The

US-based company benefits from 'personhood' under law that gives it a right to exist, and it is still operating decades after that tragic event.

It's a strange thing to admit that we've given corporations 'personhood' under law, which not only gives them a right to exist but also to use that right to destroy our environment. Tellingly, equivalent rights haven't ordinarily been given to the environment or to the protection of animals on which ecosystems – and therefore the entire economy and its people – also depend.

So, what about the Maugean skate? What about its right to exist?

The kind of fish farming affecting the Maugean skate is being met with fierce opposition. Two and a half thousand Tasmanians have written to the Federal Environment Minister complaining about it. This poor soul now has the difficult task of balancing the interests of these people with an industry worth $600 million and 5,000 jobs.

What we should be asking the Federal Environment Minister and anyone else we can get to listen is why does the ecosystem and its constituent wildlife have no right to exist when *all* lives depend on it, not just the few people who work in the salmon industry?

It doesn't have to be this way. Take Hawaii's *loko i'a* fishponds, which have existed for centuries and have been shown to increase food availability while maintaining natural function, as an example of how we can farm our oceans less disruptively. Or Budj Bim's World Heritage example of farming and producing eels for tens of thousands of years.

Animals – including humans – are both creators and destroyers, and a miracle occurs when the destructive actions of one animal (such as building a home or finding food and water) creates an opportunity for another animal to find food, water and shelter. Every time we do anything to serve our basic survival needs we destroy a little bit of the ecosystem around us, and if that action is sustainable it creates just enough waste to become life support for the other animals that live alongside us. Given enough time this forms a circular economy where everyone benefits from the creation and destruction enacted by everyone else.

This existential paradox is why no single species can create its own life support. We don't construct entire functioning ecosystems, instead we contribute to them, and this is why we depend wholly on having many other creatures coexisting alongside us.

Until quite recently the ocean was a safe haven for wildlife, meaning life-support processes could benefit us on the global scale. We were able to depend on sea creatures to absorb the side effects of our food-finding endeavours. But while you might imagine that the issue here is to protect the Maugean skate from extinction and clean up the pollution mess we've made, that's not the whole story. You can bet the argument will be made that the Maugean skate can be saved, perhaps by being relocated or a conservation plan drawn up with a set of arbitrary targets – the whole thing funded, of course, by the salmon industry to offset their impact. It's a common outcome for the agents of decline to be given the job of becoming custodians for nature, but it is pertinent to be sceptical about how much they can be trusted.

Could this be construed as subletting corporate rights to the species? Possibly. That solves nothing, though, because in truth the skate's inalienable right to exist can only be guarded by people who believe it also has a right to autonomy, to live its life the way it's meant to. It is unlikely that corporations are the right people for the job.

But what has all this got to do with me, you're thinking? Surely a Maugean skate heading towards extinction isn't going to bring about the end of humanity. This is where you're wrong. Loss of species from their natural habitat has already had a greater impact on your lifestyle and livelihood than everything you have been led to believe are the greatest threats. Yes, you read that right. It is in understanding humanity's threat to animals that we can best mitigate the effects of the most widely acknowledged environmental impacts on our lives.

For example, climate warming strongly reduces the ability of plants to photosynthesise – to turn carbon dioxide and water into the sugars that feed our planet's wildlife. However, that's only one side of the coin. It's also been established that just one-third of decline in the presence of species in an ecosystem reduces the capacity of plants to reproduce. The same magnitude of decline in plant productivity happens when the climate warms, which also threatens our food crops[39]. Further to that, when wildlife losses reach 50 per cent this starts to have an effect to rival ozone depletion, ocean acidification, elevated carbon dioxide and nutrient pollution[40]. We've been underestimating the importance of wildlife all along.

We have already lost this resilience to environmental pressures and it means we are also losing up to 30 times the earning potential nature would give us if we invested more wisely in nature restoration or 'nature-positive' outcomes[41]. Without restoration of healthy populations of species within an ecosystem every other future conservation measure will fail, economies will fail, salmon farms will fail.

Even if we have only half the local species increasing and half the skate population still exists, there is still cause for hope. Reversing the decline of the half that are suffering is society's first and simplest step to a brighter future.

Nowhere should we be more concerned about the impact we are having on wildlife than in the ocean, because this is the planet's main heat regulator; as our Sun has aged, Earth has become exposed to higher temperatures. This makes it even more important that our ocean's systems function more effectively than they did millions of years ago. Significantly it is not the mere presence of the ocean that cools Earth, it is the ecosystems and lifeforms that live in it.

About 80 per cent of the world's biomass is on land but most of that is plants. Seventy per cent of global animal biomass, meanwhile, is found in the oceans[42]. As humans are animals and we need ecosystems to survive, we're more dependent on the animal-driven components of Earth's biosphere. If we look after the animals, the plants look after themselves. These are the reasons why healthy oceans are critical to our life support.

Deep-sea mining is a bad idea because the seafloor's animals are gatekeepers for gargantuan amounts of carbon. Cool water beneath the ocean's surface mixes with the upper layers, extracting the heat energy that's continuously provided by the Sun from the atmosphere. The carbon dioxide that would allow our planet to warm beyond liveable means is trapped and stored by ocean biota. This is our major carbon storage.

Most dead animal life at the surface settles on the seabed and with three-quarters of the planet being ocean, that thick sediment layer is immense. In parts of the Atlantic this layer is a kilometre deep and was deposited at rates of about a third of a centimetre every 1,000 years. It's held there safely by a huge biomass of alien-looking creatures. Without these, deep-sea mining will kickstart ocean climate change by releasing sediment plumes the size of continents.

For me, the ocean has become one of the few places left where I can still see wondrous wildlife performing the way it should. There I can see the simple truth that nature is everything when it comes to our future survival. To understand how it is possible for humanity to have hope for the future, we must look under the sea.

The Distinctly Unsinister Sea Urchin

It was half an hour before sunrise. Beams of orange sunlight were starting to glow in the humid air, stencilled onto the deep blue of dawn. I watched as wisps of cloud were slowly revealed, hugging the summit of Mount Tambora – the active volcano on Sumbawa, Indonesia. In the peace of the morning it was almost hard to believe that the slumbering Mt Tambora was responsible for the largest eruption in recorded human history.

On the day of its last eruption on 10th April 1815, the inestimable explosive power of nature reduced the mountain to two-thirds of its size. A plume of super-heated ash flowed to the sea at 200 km/h, destroying all life in its path. Sixty- to ninety-thousand people died in the ensuing disaster, many more than in the aftermath of Vesuvius and 10 times more than Krakatoa. Almost all the island's inhabitants were killed, and the event led to the infamous 'year without summer'.

For several seasons the world's climate was altered, causing widespread crop failure and famine and pushing food costs up to eight times higher than they had been before the eruption. Reportedly, 19-year-old Mary Shelley – on summer holiday in Switzerland with Lord Byron and other writers – was driven indoors by relentless rain and fog. In their confinement they challenged each other to write the scariest story they could imagine. Legend has it that this is where Shelley embarked on her most famous novel, *Frankenstein* – the story of a reconstructed man brought back to life.

Indonesia is one of the most volcanically active places in the world. Two hundred years after the Mt Tambora eruption and the Sumbawa Regency is inhabited by half a million people; its ecosystem has been stitched back together by wildlife, its soul reignited by the Sun and nature flourishes as it has always done. Life began in the sea, and the power of nature to restore is perhaps nowhere better demonstrated than here.

For the local people, coexistence with nature has been part of their culture for thousands of years. As the world's largest remaining nation of sea nomads, the Sama Bajau understand the enduring resilience of nature. *Isa' lau itu niak Saloh*, they say. If not today, tomorrow is still there[43]. Their cultural identity is built on knowing why shifting seasons, changes in climate, natural disasters, wildlife and the ocean are part of an all-encompassing and abiding force that maintains the health and prosperity of our planet.

Throughout Indonesia people live in the shadow of volcanoes, settling close to the sea at the base of these mountains. It is through this combination of fire and water that they understand the balance of the ecosystem that sustains them. Just north-west of Komodo is an almost perfectly circular island some 2,000 m high containing Sangeang Api volcano with its twin craters. Rain-eroded peaks and gullies tower over slopes clothed in monsoon forest. The sea at the bottom of these dark green drapes is festooned by gardens of feathery black coral growing over warm sand and shimmering curtains of volcanic gases bubbling skyward. Pygmy seahorses live here, the males of these small, tulip-pouted fish kept perpetually pregnant by a female.

Away from the margins of the coast, false killer whales cavort and hundreds of sperm whales dive to dizzying depths in search of squid. The whales' complex cetacean societies include lifelong maternal bonds and post-menopausal lives where grandmothers look after offspring. Their ancient culture exists among mysterious patterns of migration that send some of them around the whole planet.

These extraordinary creatures – among the largest that have ever lived on Earth – store complex histories of sound and magnetism in their brains. By connecting this past wisdom to a deep understanding of ocean climate and currents, they can use this knowledge to chart their movements and follow patterns of distribution and abundance in their prey. So long as there is cetacean society these collective learnings can be shared and handed down over generations, giving future newborns the best chance of survival. Communication by language favours larger, more mobile animals, so our world is dependent on the surviving culture of these cetaceans that do the all-important job of balancing the world's ocean climate.

Closer to the surface, at the volcanic coalface, the strangest creatures live. These are more unusual than anywhere else on the planet. They fulfill a wide variety of peculiar roles and connect obliquely with each other's lives, tapping into the erstwhile riches of a land ignited by fire, rained upon by black ash and smouldering with potential to both collapse and create whole civilisations.

The sediment below Sangeang Api is black sand, like the best organic mulch. Thousands of years of volcanic spoil rich in magnesium and potassium nutrients still drains off the mountainside. This charred moonscape dotted with dusty corals might appear barren at first but look closer and you can see life forming as it would have done hundreds of millions of years before.

Parades of incandescent sea urchins pass by. You'll see fire urchins glowing like charcoal embers, edged in cobalt blue with radiating shafts of bright orange and yellow. Then there are the less flamboyant and hairy heart urchins – some of them occasionally stumbling across the sand uncannily fast on a thousand legs.

Sea urchins are incredibly biologically and culturally sophisticated. Scientists think some species could be practically immortal as they show no signs of aging. Immortality does not mean they can live forever, rather they just do not age. But no animal is in absolute control of its destiny and probability dictates that they will be eaten at some point or a rock will collapse and crush them.

Studies of sea urchins further reveal their fascinating sophistication and the lengths they can go to in building a habitable world. Scientists have found light receptors in their feet, suggesting that the entire urchin acts like a huge compound eye. Urchins can also sense minute chemical gradients in seawater to find food and send and respond to signals to each other that alter their behaviour. They can then form 'herds' to forage together or flee predators. Given their enormous sensory ability and long lives, the development of a culture of cooperation is unsurprising.

There are few animals more maligned than a sea urchin. Helen Sullivan writes in *The Guardian* that 'sea urchins are as sinister as they appear' and 'have a darkness'[44]. Sullivan only reports what she has been told, which is that they have laid waste to kelp beds. The reality is quite different.

This perception of darkness comes from fear of the unknown. Our society's disconnect from nature can propagate an almost supernatural hatred of some species, when instead we could bask in the mysteries they contain and marvel at how little we know about them.

Where I live in Melbourne researchers are literally smashing these ancient, highly sensory and marvellous urchins with hammers to control their numbers. Yet these ancient echinoderms are also found to be abundant as fossils embedded in soft sediment that was laid down when the region was volcanic five to six million years ago – proof that they've always engineered our living landscapes.

The seafloor in one of our most important marine sanctuaries is littered with carcasses, despite there being little to no evidence these native urchins are in plague proportions here. Native sea stars and sea urchins have endured precipitous population declines over the past 10 years and it is only in the places we depend on for food and coastline stability that they are becoming more abundant than we would like them to be. There it is – once again humanity acts on the arrogant perception that we have a right to determine and control the existence of another species.

Urchins are essential for reef health. At the base of the reef they dig spherical holes in hard rock with their teeth, leaving grind marks on the inner surface. The long-lived urchins, ones with the greatest knowledge of how to survive, build themselves a crevice in which to hide. When conditions aren't favourable they can shut down and enter a 'zombie' state where their internal organs shrivel up and they wait until conditions are better. Then they reform their insides and start living again. Even the young ones are essential, as abalone (another important herbivore in our aquatic ecosystems) seek shelter under the spines of sea urchins.

Abalone and urchins feed on the same plants, and as a result overfishing of abalone has caused urchins to become more abundant to fill that gap. Sea urchin abundance has long been known to depend on predators such as lobsters and fish, but rock lobster is good eating and it's legal for any one of the five million people who live in Melbourne to catch a few lobsters with a licence. Early sailors used to be able to reach down and pick them up in the shallows – at low tide lots of antennae could be seen sticking out of the water. Now that sea urchins are being removed, there are none.

On the Great Barrier Reef, a sea star called crown-of-thorns (a relative of the sea urchin) is thought to be destroying reef and suppressing marine recovery, especially after cyclones. Scientists are designing AI-driven machines to kill them. In other cases, working teams go out with vinegar in huge syringe-like apparatus and inject them with the poison. Given the scale of the problem, this is never going to be a wholesale solution. Meanwhile, the causes of this imbalance are known.

There are more than a few suspects in the murder of the Reef. Firstly, the species most likely to have kept the sea star under control, the giant triton sea snail, was collected as a curio and sold in markets in the past. Secondly, the Reef – like all coastal areas – has fertiliser residue poured into it from farmland runoff. Finally, as the Australian Institute of Marine Science recently discovered, there is a very strong relationship between the abundance of crown-of-thorns starfish and removal of predatory fish[45]. It seems the crown-of-thorns has been wrongly accused.

Such discoveries seem obvious, yet the need to prove this beyond all doubt seems to lead to delays in addressing the cause and rescinding the death penalty for these vital creatures. Instead, restocking the ecosystem with wildlife and reducing our threats to it is the most effective and scalable way to protect ourselves from the more sinister effects of ecosystem collapse.

We visited a coral reef off the southern end of Taveuni in Fiji near a resort run by lifelong dive guides and resort owners Terri and Allan Gortan. Allan tells me that after Cyclone Winston in 2016 the site was infested by crown-of-thorns starfish, so much so that they considered it to be worthless to return to the world-famous dive site. A few years later a US group visited and begged to see it. Allan reluctantly agreed, warning them it would likely be barren. But on arrival they discovered very few crown-of-thorns and an abundance of recovering coral reef. While it's anecdotal, Allan believes that the starfish were somehow responsible for the recovery.

Recent research suggests he may have been right. Monique Webb at the University of Sydney has found that young starfish, which begin life feeding on algae, stay herbivores in the presence of adults rather than progressing to consuming coral. It's all to do with how they smell[46]. Remove the smelly adults and more juveniles grow into their large coral-eating form. While researchers postulate the use of artificial chemicals to restore balance, they freely admit that outbreaks of adults appear to happen only where the starfish's natural predators are absent.

I don't know any scientist who has ever considered the possibility that outbreaks of crown-of-thorns starfish might have some benefit. It makes sense because, like sea urchins, crown-of-thorns starfish have always been essential to reef recovery. Reef ecosystems are a bit different to land where vegetation tends to dominate. On reefs, coral, sponges, hydroids and diatoms are all types of animals. Crown-of-thorn starfish graze reefs like turtles graze seagrass. The quickest way to restore a functioning ecosystem is to do the equivalent of pruning, ploughing, harvesting and sowing, which is what the crown-of-thorns are doing. We see the same when we introduce large animals like beavers or bison back into temperate grasslands – that grassland recovers faster than ever. There is no reason why coral reefs are any different.

While reef catastrophes have always happened, they are occurring far more regularly than they were a few hundred million years ago. Knee-jerk actions to remove crown-of-thorns starfish and urchins now could easily negate a process that would aid in their recovery. Jeremy Day from Wollongong University has revealed that one species of sea urchin from south-east Australia isn't even a strict herbivore[47]. Mostly, there is little difference in their abundance and body condition within or outside the kelp beds. The problem does not lie in the abundance of urchins or starfish at all, but in the overfishing of their predators such as sharks and, in the case of crown-of-thorns starfish, even giant triton snails. Again, it is not about taking out more animals but trying to rebuild the delicately balanced ecosystem we've lost.

Allan must now ask visitors not to kill the crown-of-thorns starfish on the resort's house reef. Back in Melbourne, I attend an event run by a senior government scientist where members of the public stand up and beg to be allowed to kill all the urchins that they think are destroying their marine ecosystems. I see scientists make these ancient echinoderms into mortal enemies and wonder what's gone wrong.

For as long as there have been humans, urchins have wandered the oceans. Whenever there were volcanic eruptions, ice-ages or other catastrophic instability in the biosphere, echinoderms would pick up the pieces and begin rebuilding. Their superbly successful adaptations – that are inconvenient to us and make them a pest in our eyes – have allowed them to restore, rebuild and maintain functional ecosystems for all that time.

Why do we think of them as a pest? In disturbed ecosystems dominated by urchins there is an excess of nutrient that manifests as too much plant and seaweed growth. This benefits a larger number of smaller urchins that are less nutritious to predators and that forage intensively. Over time they start to restore balance, reducing the excess plant nutrient. As a result fewer young urchins survive, leading to aggregations of longer-lived and larger urchins and a greater diversity of plant life. Those urchins begin to understand where the remaining food is found as they have lived long enough to store this knowledge in their brains. They also become more nutritious, which attracts predatory fish and provides surplus food to people.

There are few places where we can watch and learn from sea urchins performing their daily rituals. These daily habits are lessons in why their peaceful offering is important in the increasingly volatile world that we have made for ourselves. Our beliefs about sea urchins have become as degraded as the ecosystems we live in. With human-induced nutrient overload now affecting almost every reef in the world, we are turning our coasts into a place that need sea urchins more than ever to secure our immediate future. Their unique survival ability will ensure there is stock left to rebalance the catastrophic eutrophication of the reefs on which our very lives depend.

We tend to use science to justify culling urchins because in our self-involved arrogance we think they are bad. Instead, we should base our actions around the philosophy that urchins are good and use science to understand how we can rebuild balance around them. A lot of this has to do with what we believe is a good-looking ecosystem, but in that sense we are not very wise.

It all comes down to the perception of what we think is a 'healthy' reef. Most of us are only familiar with a particular type of coral reef that's dense, lush and highly structured. However, in places where there is naturally high nutrient input the system is very different. The former is akin to the natural opulence of the Amazon rainforest, while the latter looks more like the African savannah. Both landscapes are equally important, and though the savannah looks flat and empty, to the curious eye it is full of animal life. Perhaps this is another reason why we are so keen to dismiss these landscapes as destroyed and the creatures that inhabit them as destructive – because they do not conform to our ideal of what a healthy, living reef looks like.

So where are the living creatures of these savannah-like reefs? Many of them – like the colourful carnivorous nudibranch – are hiding out. Go out at night and sea urchins appear in droves. Significantly, this diversity of animal life in plain view is a direct consequence of the naturally high nutrient input.

Why don't these nutrient-enriched ecosystems collapse like those we see at the Great Barrier Reef or throughout our own coastlines? One answer lies at the mouth of the Mayalibit Bay in Raja Ampat. We recently dived here and it is easy to believe the seafloor may have never been seen by people before. There is a sandy slope where freshwater mixes with tidal salt water and enters the sea. Here, finely ground material washing downriver from the plants and limestone rocks above has had the chance to settle for up to 25 million years. The substrate, like on the edges of the volcanoes, is deep and if you touch the bottom you risk clouding your experience with an eruption of silt.

Ten metres below there are encrusting corals, sponges, delicate tunicates, a patchwork mosaic of bare sand and dense, spindly coral fingers like spaghetti. In half an hour we counted a remarkable 14 species of delicate nudibranch, including the 'Marilyn Monroe' with its creamy overcoat covered in dark spots and purple frilled dress that blows upwards in the current. There were pipehorses, bright yellow sand gobies living in holes with their cleaner shrimp tenants, bizarre polka-dotted juvenile barramundi cod, fragile porcelain crabs and, of course, a variety of peculiar-looking sea urchins.

The living latticework of creatures working together has reached a point where almost all the surplus nutrient is consumed. This is the investment of wildlife – each destroying the ecosystem a little so others can create a home, stabilising a highly dynamic river mouth and turning it into the Garden of Eden. This makes this landscape an extraordinarily fragile and rare example of a balanced, working ecosystem. Nothing else compares except perhaps the deep sea – a place we're about to disturb on a massive scale with equally catastrophic consequences for nutrients and ocean climate.

Fortunately, our attempts at controlling sea urchins are feeble compared to the creatures' power to alter whole ecosystems for the better. Some scientists won't like me saying this but it's true. Where sea urchins and sea stars become a real inconvenience or where they have been introduced and are 'pests', we stand no chance against their advances in the long term. The only thing we can do is buy ourselves time to rebuild the structure of our coastal ecosystems by reintroducing what's been lost and restoring the balance that these places were built on in the first place. But these moves must be carefully moderated to ensure we don't take things too far.

According to the Merriam Webster dictionary, Manu Leumann suggested that the word 'urchin' may have originated in soldiers' speech, with *ērícius/ēricius* originally alluding not to a literal hedgehog, but rather to an obstacle with sharpened ends used in fortifications. This is a better way to think about urchins – as defenders, not destroyers, of our ecosystems.

Even if we feel the need to control their numbers in some places, isn't it better to do this with a more informed belief systems based on respect and care? After all, urchins have become a species most likely to survive and that gift of immortality surely makes them a very powerful ally for our future.

Though no-one speaks on behalf of the lowly but immortal sea urchin and generally we know little about them, they are not creatures to be feared. What they do for us is inestimable. They clean up our mess and help limit the amount of pollution reaching our seas, leaving just enough surplus for larger animals to make a living but not so much as to collapse ocean climate. Our cultures, knowledge and information systems stay relevant because of this action and our very lives depend on it. If we are to learn one lesson from the distinctly unsinister sea urchin it's to not fear the unknown, but rather to reflect on how much we don't know.

When we degrade ecosystems beyond our own species' individual capacity to repair, there are countless animals ready to step in and restore balance. Some of these we take for granted. Many more we view as pests, when in fact they might be exactly what we need to reverse the consequences of our own species' misbehaviour.

Whale Sharks and the Sama-Bajau

The morning air was thin, cool and silent. The wind had dropped overnight, and the tranquillity of dawn subdued our excitement to whispers. We were all sitting in a circle on board a small boat preparing our masks, snorkels, hoods, cameras and nerves. We were about to enter pitch-black water knowing that within 20 metres of us swam several of the largest animals on Earth.

We had cruised overnight for about 50 miles to arrive in Saleh Bay, a 2,000 km² cove separated from the Flores Sea by two narrow channels. It may be one of the few places in Eastern Indonesia with resident populations of whale sharks. It is close to 300 m at its deepest and gathers nutrients from the surrounding volcano sides that turn it green and encourage plankton.

Our first glimpse of a whale shark through the murk was of one hanging almost vertically in the water. Brilliant blue dots also glistened in the depths; these were sapphire copepods, the crystalline structures in their paper-thin bodies reflecting blue light under the gloomiest of conditions. The copepods can dissolve their reflection in an instant and disappear like a ghost. Along with the whale sharks, these watery jewels migrate to the surface at night when the local people also fish. Humans and whale sharks share this home and have become accustomed to living and fishing together for thousands of years.

In this special place whale sharks live in harmony with the Sama-Bajau – the predominant cultural group of Eastern Indonesia's marine regions. The people of this culture have plied their trade fishing here for 12,000 to 15,000 years and they have a variety of customs related to whale sharks. Hunting whale sharks is forbidden in some clans as they are considered sacred animals and are believed to protect fishers from harm. The same applies to whales and dolphins, too, which is why there are very few indigenous villages in the whole region that traditionally hunt any such creatures.

This relationship has served the Sama-Bajau well, as whale sharks create ideal conditions for their fisheries to thrive. As we watch on, several 20-tonne whale sharks congregate around a *bagan* – a type of fishing platform that uses light to attract baitfish at night. Although *bagan* fishing is a recent phenomenon, it's been customary for these fishermen to protect the sharks. They give back a portion of their catch to their friends underwater and the sharks have grown accustomed to this relationship. Local fishers had completed hauling their net from below the *bagan* about an hour before we arrived.

As the sun rose above the nearby volcanic peak of Mount Tambora, the first shafts of golden light revealed the full length of the sharks. Saleh Bay is mostly home to male sharks and of the five we could see we estimated the largest to be about eight metres long, which would make it a mature male. No-one knows very much about the females, where they go or how the males move from here to meet them. Newborn whale sharks, which are only about 30 cm long, have hardly ever been seen. Despite our best efforts the largest animals on Earth will remain a mystery.

The water was green and peppered full of life flowing past in a moderate current, which meant we had to keep snorkelling upstream like weird neoprene fish. We saw chains of salps, a gelatinous filter-feeding organism with prominent brown nucleus. There were also lots of small jellyfish and microscopic plankton everywhere. Aboard the largest of the whale sharks there was one lonely remora (suckerfish). Another smaller one was living on the tail of a second whale shark.

At the expense of less mature whale sharks, the largest shark fed almost continuously for the four hours we were there, still in that upright position. Unsurprisingly, vertical feeding uses less energy than pushing a 20-tonne body through the water for hours. The crew on the *bagan* were dishing out krill from buckets, literally tipping it into the hungry mouths of the sharks. It was like watching baby birds begging for food in the nest. As Westerners we are prone to worrying about the ethics of this type of practice, but who are we to question the relationship between two animals that have shared an ecosystem for thousands of years?

As krill scraps are added to the surface, the sharks emerge with their snouts then sink with mouths open. The subsequent rush of water into the cavernous mouth is followed by a flourishing of the gills. Up to 600 cubic metres of water an hour can be consumed by a shark and this has to be filtered by the gill rakers before the food scraps are swallowed. There are rows of tiny black dots below the eye called ampullae of Lorenzini – electroreceptors. For much of the time whale sharks feed in the deep, dark ocean and these no doubt help them detect patches of food more effectively than their tiny eyes.

A marine-mammal biologist friend once said they were unimpressed by whale sharks, while another remarked how few facts about them are exciting apart from their size. It's true. We know almost nothing about them and I doubt that we'll discover anything extraordinary enough to compete on National Geographic's *Shark Week*. But for me there is something extraordinary in that which is considered unremarkable.

Whale sharks are slow-moving filter feeders that don't even forage in the highest food gradients. They leave that to blue whales, which might occupy areas we traditionally call 'biodiversity hotspots'. The sharks have a thick blubber layer and a very slow disposition, perfectly suited to living on the margins where nutrient is scarcer and resources limited. It's a larger area, though, and overall has enough potential to tip an ecosystem out of balance when it is without its regulating life forms. Whale sharks are essential to maintaining the balance and keeping a whole food chain intact and this is especially important for local livelihoods. The Sama-Bajau may not have scientific data to back this up but they know it through their age-old cultural wisdom that has come about through their relationship with these creatures.

There is no other animal so acutely evolved and capable of doing this here than whale sharks. This is revealed when the sharks' spotted patterns take on a different meaning at night. They blend in perfectly under faint bioluminescence and, contrary to popular opinion, aren't camouflaged to avoid predators. Instead, these stealthy behemoths use that disguise to quietly creep up on and among their prey. They make themselves appear part of the plankton and spend their lives vacuuming it up in vast quantities.

The glacially slow actions of whale sharks belie a greater truth in that they embrace robust, animal-led systems that play an enormously powerful role in the evolution of our own survival and the regulation of Earth's life support.

Scientists from Charles Sturt University interviewed Sama-Bajau fishers about their ocean friends, some of whom told the interviewers 'our ancestors forbid us to catch whale sharks'. An elder from the village of Sulamu in West Timor said that the shark is guarded by a spirit who keeps fishermen safe. There are stories of fishers from nearby Kera Island who held onto the fin of a whale shark that saved them from drowning and towed them to safety[48].

Sama-Bajau people believe they are more than just tied to the soul of the sea, they are connected to it by birth and animals are members of their family. The placenta represents their twin *Tamoni* returned to the ocean to grow like a human before assuming its identity as the spirit of a sea animal. This deep understanding the Sama-Bajau have of their connection to both living and non-living parts of the ocean comes from their ancestors.

Their society follows a process that has been passed down through generations. On the southern tip of the island of Sulawesi in Indonesia are a chain of islands that arc south into the Banda Sea and are home to several Sama-Bajau villages. The island of Tomea is the gateway to the remote and extraordinary Wakatobi Resort. Resort founder and lifelong conservationist Lorenz Mäder describes to us how the Sama-Bajau would settle an area, deplete its fish and move on after 50 to 60 years.

Though effective, traditional practice isn't permanently sustainable because – like everything – it's influenced by outside changes. Human populations have increased, and there's overfishing, marine pollution and climate change to contend with these days. But the Sama-Bajau's solid appreciation of nature may be their saving grace. This belief in the significance of the ecosystems around them makes them more resilient to the whims of the sea and a changing economy. They are more likely to accept and stand by changes to their traditional practices that preserve their food security and way of life because they are still connected to nature every day. If they resist these changes, that connection falters.

Lorenz and the Wakatobi Resort are partly responsible for the local people being able to easily navigate this transition. The success of this relationship stands as an example of what we can all do to help ourselves move to a new nature-based economy.

When Lorenz first arrived, the coral reef was mostly gone and there were very few fish at all. From the outset, he asked the village heads if he could lease dive sites from the islanders. In return for halting fishing, they would be paid a premium above the value of the fish they would have sold in the market.

Within a matter of years fish started to return to the areas of reef between the protected dive sites. The local fishers were reporting more and larger fish. In the years that followed the resort's success grew, and today 17 villages are involved in this partnership. Now hundreds of species of resident reef fish swarm over the healthy coral and turtles cruise by with giant trevally, batfish and reams of red-toothed triggerfish.

> **LESSON 8:** Watching animals allows us to interpret complex processes and make smart decisions to chart a course to a better future.

The arrival of creatures like whale sharks has always signified seasons of plenty. Perhaps they are revered not because of their intrinsic importance to the ecosystem here, but because their unfathomable spirit connects living people to nature. Without the guidance of animal behaviour, humans will not be able to decipher the map that leads us into tomorrow, next month, next year or the next 100 years.

For the Sama-Bajau, thousands of years of ecosystem knowledge has been passed down orally through stories that connect an uncanny ability to navigate by the night sky with the comings and goings of wildlife, seasonal weather patterns and even natural disasters. In Western society we lament the way the human brain isn't used to full capacity, evidenced in the fact that when we need to navigate we reach for the soulless, desolate directions of Google Maps. In 2006 the US Navy even stopped teaching celestial navigation but is soon to resume it given the growing threat to our satellites and GPS.

For the Sama-Bajau, their connection to the whole seascape is a matter of life and death. Survival depends on filling their brain the natural way – with information about wildlife. The seasonal amplification of food systems by whale sharks meant the Sama-Bajau could read the ocean, congregating in the richest feeding grounds by using astronomical navigation. It is humbling to think that without nature we become nothing – with nature we become everything.

Whale sharks have been around for 245 million years, hominins (humans) for just 2 million. Where evolution is concerned, there is no effective starting point to our connection with nature. We adulterate such ideas by imposing abstract notions of evolution. For instance, what even is a 'species'? There wasn't a single moment when humans became the species *Homo sapiens*. We evolved from another species, *Homo heidelbergensis*. Our culture, behaviour and connection to nature and wildlife is a continuum that spans timelines that match the slow, indefatigable march of culture and evolution. Our relationship with nature is timeless, as the Sama-Bajau know, but Western science so often reduces this to a lifetime.

It's not that these people researched or learned how whale sharks mattered, they just came to know through the passing of culture, traditions and knowledge through many generations. Their natural philosophy and beliefs evolved as a way for patterns of endeavour to exist for as long as possible. Throughout history, humans became among the likeliest species to survive because of an intrinsic respect for nature – something that the Sama-Bajau still maintain. Today, the challenge faced by Western society is to learn about and develop this respectful relationship once more, but to do this we must understand why it's needed. This age-old bond between the Sama-Bajau and whale sharks is a clear lesson in how understanding can lead to respect and map a secure future for us on this planet.

The size, demeanour and behaviour of whale sharks in a nutrient-poor world means they have a disproportionate impact on our lives. Like manta rays and blue whales they transfer, amplify and concentrate nutrients. We cannot even begin to estimate the intensity and precision through which they do this. Nor can we fathom how they navigate, how these experiences are passed down through generations and how they occupy specific places as they have done practically forever. Their sustained presence and actions soften weather extremes and condition and fertilise the ocean, providing food that we take full advantage of.

The intensely nutrient-poor waters in parts of Eastern Indonesia would remain so if it weren't for whale sharks. Because of their size and mobility, they can reach places where the sun's energy is captured fastest by marine algae (plants). By consuming vast quantities of the extraordinarily energy-rich plankton that feeds on the algae they kick-start amplification ecology. This kind of ecology is when the sharks continuously recycle and maintain those processes throughout the season. This practice produces waste that vast congregations of other animals benefit from. Whale sharks don't simply maintain balance in our ocean ecosystems but are instead an ecological vortex, sucking in other creatures to a whirlwind of processes that season the ecosystem and turn it from bland to flavoursome.

When scaled up to consider there are about 150,000 whale sharks in the Indo–Pacific (maybe as many as half a million globally, before declines) their processes have consequences for our global climate, too. And the scale of impact from whale sharks isn't only in immediate time and place, it has been going on for hundreds of millions of years; long before human civilisation, whale sharks were laying the foundations for our evolution.

Whale sharks forage almost continuously, coming to the surface at night and diving deeper during the day. According to the Shark Institute, an average six-metre whale shark filters 2.8 kg of plankton per hour for eight hours a day from 600 cubic metres of water per hour. At last count, 99 distinct whale sharks had been individually identified in Saleh Bay. Let's assume, conservatively, there are 200 whale sharks in total. These animals are filtering 328.5 million cubic metres of seawater annually. This is a staggering 0.15 per cent of the entire volume of Saleh Bay.

This might not sound much but we're land dwellers. We think in two dimensions because our impact is on the land's surface and barely a metre below. The three-dimensional ocean has a lot of relatively empty space and whale sharks aren't confined to a thin layer. They have a massive influence on ecosystems because they can move and consume food where it is most highly concentrated – over hundreds or even thousands of metres vertically. This contribution of hundreds of thousands of whale sharks moving up and down to transfer nutrients from the deep back to the sunlit surface where we fish and feed is one of nature's most extraordinary services. Without filter-feeding whale sharks we have no way of reaching our food. Worse still, we end up living in algae-ridden, carbon-releasing dead zones with no fish at all.

If we lose whale sharks this will threaten food security and climate for billions of people. Yet, somehow, this vital piece of information is missing from primary research directed at saving wildlife. One of the more important findings from satellite tracking whale sharks is that tags often 'switch off' in high-density shipping channels. Researchers suspect, quite logically, that whale sharks are being struck and killed by ships.

But what difference does it make to decisions about our future to know this? To survive the next 100 years comes down to answering this question, then working out how we can change hearts and minds to move shipping channels, slow down ships or avoid killing sharks.

Until whale sharks are given a right to survive and are recognised as imperative to the survival of human life, there will be no chance of changing our own behaviour. That belief system needs to permeate all scientific discourse if we are to have any hope of mapping out a successful future for our species.

Humble Hermit Crabs

We arrive at the village of Ambabee in Raja Ampat at dusk and set off from there to our destination. It's a short walk past piled-up coconut husks to the base of the hills we're seeking. The jagged gothic peaks that rise above soft sand were made from the skeletons of 150-million-year-old coral, forged underwater by animals and buckled upwards after continental plates collided about 25 million years ago.

It is here among the ancient reefs of their ancestors that coconut crabs seek refuge. Coconut crabs don't look like an animal that can easily climb over sharp rocks, seeming more at home on the trunk of a tree. As they clamber about, their shells connect with the limestone and give off a delightful sound like expensive China clinking together at a tea party. These pottery shells are shiny, imbued with deep reds and blues, and their long antennae are constantly tasting the air.

Coconut crabs are found throughout the Indian Ocean and Pacific Ocean but are a type of hermit crab that's lost its ability to survive in the seas. Our guide for this trip, Florens, is a local West Papuan chosen by *Conservation International* as custodian to lead a project protecting the crabs. Over the years tourists have become more and more intent on eating them and it's led to their rapid decline and local extinction. The island provides a much-needed sanctuary.

I'm eager to discover if there are any traditional rights that protect the crabs and what benefits there are to saving these creatures. 'There are none' comes the response to the former, and 'they will attract tourism dollars' is the answer to the latter. It's a start, I suppose.

Raja Ampat has had a fragmented history of human occupation. Rock art in Misool has been dated from several thousand years and there were, until recently, pottery artefacts accompanying grinning white teeth from skeletons in low-hanging caves, possibly placed there at a time when the sea levels were lower and a land bridge formed between the islands and mainland. Until quite recently most settlement was thought to be temporary. Fishers, possibly from the Sama-Bajau, would pass through and set up camp, but fresh water is hard to come by here. Permanent settlement by West Papuans is only very recent, which might explain the lack of customary protection for the crabs.

For now a bid to maintain tourism to the area is as good a reason as any to conserve coconut crabs and we are happy to support it. Local people here are among some of the poorest on Earth with an average salary of only US$150 a month. Our guides have fixed coconuts to the limestone wall so the crabs clamber into view, and as we watch these fascinating animals I am left wondering how much more needs to be done before they realise the true importance of these crabs for the villagers' own future. What happens when there are no tourists?

For thousands of years humans elsewhere tolerated the crabs and they earned a place in traditional culture and knowledge. But an animal that is so good to eat cannot possibly survive human occupation unless there are very good reasons for it doing so. One of these reasons is the chance to be poisoned. There is some evidence that consumption of certain fruits can lace the coconut crab's flesh with toxins that can cause illness, heart attack and death. This belief alone may have been the key to the animal's survival for so long.

It is unfortunate that tourists come along to eat the crabs occasionally, and the irony isn't lost on us that the very act of tourism that will save the crabs is also the reason they are going extinct.

Humans first arrived on the remote islands of Fiji about 5,000 years ago and the Lapitas of the Neolithic age are thought to have benefited from their relationship with hermit crabs. These animals come in all shapes and sizes and have had a remarkable effect on human settlement, habitation and survival. Katherine Szabo from the University of Wollongong observed that two-thirds of the shells in an ancient midden (essentially a waste dump for their discarded shellfish shells) on the island of Ugaga had been exchanged by hermit crabs. Szabo suggests that the abundant crabs were providing a cleaning service consuming domestic waste[49].

Hermit crabs live in a world of smell. They forage mostly in the dark using sensitive sense organs to detect rotting material a long distance away. If you walk along the beaches in Fiji and turn over fallen coconuts, it's not uncommon to see dozens of them crowded together in a feast.

The grouping behaviour of hermit crabs is critical for nutrient processes. The crabs themselves benefit from a constant turnover of resources. In the same way as all animals build ecosystems, they themselves engineer the processes that support new growth and feeding opportunities. Island sand habitats are largely devoid of nutrient patches, so congregating hermit crabs would collectively fulfill the function of larger animals that are absent on remote islands by collecting, transferring and amplifying nutrients into convenient patches so that larger animals can forge a living. For the most part, however, there were no large animals able to survive on islands until humans came along. The small hermit crabs must have laid the foundations for their own evolution into the biggest coconut crabs.

As shown in a Taiwanese study, hermit crabs have evolved to exploit gaps in the ecosystem by forming non-competitive alliances. On islands with coconut crabs, numerous other hermit crabs and land crabs live side by side. This is where human–crab interactions are important.

> **LESSON 9**: Earth is still ruled by the animal kingdom. Be humble.

Like all animals, hermit crabs have coexisted with each other and humans by fostering the building of shared ecosystems. The only reason island settlers could survive was because the crabs had accumulated enough nutrient in the ecosystem to support the arrival of these large-bodied animals and could adapt their behaviour to maintain this support even through the inevitable disruption to the island's ecology that the arrival of humans caused.

While it's hard to separate the impact of different species, hermit crabs are a huge part of coastal biomass. The small ones occupy the smallest areas and pile up nutrients in tiny patches, spreading them widely. The larger the crab, the more concentrated the nutrient patches and the greater the impact on our survival. Their nutrient-amplification role helps promote fish abundance as rainfall washes the nutrient down channels into the sea. These systems are finely tuned, fragile and localised because coconut crabs are resident animals with territories that are mostly less than three hectares.

It takes 40 to 50 years for coconut crabs to grow to full size, which is about a metre from claw-tip to claw-tip. These creatures evolved from smaller hermit crabs about four million years ago. Our species only evolved 200,000 years ago and we couldn't cross open ocean until very late in our society's development. By the time people and coconut crabs found each other, the crabs had been among the largest animals to ever inhabit these global outposts for millions of years.

Coconut crabs are unlike any other animal on Earth. Being among the largest and heaviest wildlife on tropical islands there is no doubt of their significance as the animal most closely associated with our needs. And like any important animal that lives alongside people, a strong belief in their significance is essential to ensure the survival of both them and humanity. The plight of coconut crabs in Raja Ampat and Fiji where they are facing extinction is either because there is no customary protection or previous safeguarding has broken down as ecosystems and societies have simultaneously fragmented and degraded.

As always, we place ourselves at the centre of importance and think the crabs depend on us when it's really the other way around. There is our lesson in humility to be learned from these hermits. When scientists say the crabs are important for nutrient processes, that is only part of the story when, in fact, there is a greater and more fundamental truth at play. These crabs shape entire ecosystem processes as the nutrient drivers and engineers of life support. It would be more accurate to say that we risk losing the basis for people's survival if we don't maintain a coconut crab population.

The protection of coconut crabs at Ambabee might be the last hope for humans – we don't know. Though crabs can live on and might eventually recolonise neighbouring islands (even if it takes tens of thousands of years more) that means humans will have to wait and then adapt over these thousands of years to a new way of surviving, one without these giant hermit crabs whose ancestors helped our ancestors through the hardships of living among a disrupted ecology.

On these tropical islands it's the hermit crabs that rule and humans may just be temporary visitors. Our powerful connection to what may seem the most insignificant of animals is often too subtle for us to notice. As we've lost our cultural connection to nature, it's more important than ever that we learn a lesson in humility from these creatures. It is humbling to know that animals are our strongest ally in the fight to recover or maintain life supporting ecosystems.

Grappling with Gropers

Last year a young man illegally speared a blue groper in Sydney. That might not have been enough to cause a stir, but what this feckless fisher didn't realise was that this was not just any blue groper; this was Gus, and he was loved by everyone.

Gus (may he rest in peace) was an ecosystem engineer. Gropers grow slowly, reaching maturity after 20 years or so. Living for perhaps 50 to 70 years or more they can reach sizes of a metre and weigh 18 kg – that's about the weight of a clouded leopard, roe deer, gazelle or baboon. Gus lived in an area no bigger than 10 football fields and ate 10 per cent of his body weight each day. About 25 of Gus could consume 200,000 sea urchins every year, keeping their numbers in balance and maintaining seaweed diversity. It's estimated that 18 million sea urchins might be undermining Tasmania's east-coast fisheries. It would only take 2,500 blue gropers to keep those under control.

The killing of Gus sent shockwaves through a community that loved him enough to give him an identity and name. As the community already considered themselves guardians of Gus and those like him, Gus's untimely demise prompted New South Wales Fisheries to implement a 12-month ban on any form of fishing blue groper.

The Minister for Regional and Western New South Wales Tara Moriarty even recognised the relationship between Gus and the people in Parliament, saying,

> This is of concern for the entire community: for people who fish and for people who are interested in protecting our marine environment. The behaviour of this species of fish results in people forming particular connections with it in some parts of New South Wales.

The recreational fishing sector was immediately up in arms. Steve Starling in an article titled 'Blue Murder' for *Boatsales.com.au* cited 'zero scientific evidence' for the ban, claiming that 'fisheries management decisions should be based on good science, backed by research – not on emotion, or the projection of human feelings onto the natural world'[50].

Sorry to disagree with you Steve, but there is, in fact, clear evidence of why we need large fish for coastal ecosystem integrity. Moreover, community beliefs in the value of Gus and fish like him are of overriding importance irrespective of scientific research. The consequences of overfishing on coastal livelihoods are beginning to have an impact both on inspiring more consideration of rights for nature and on what most people know they need and want. Put simply, we're starting to realise that we need animals like Gus and that they are all we have left if we are to survive the next 100 years.

The incident with Gus proves that the fishing industry's belief in the importance of fish life is not yet strong enough to act in their own self-interest, let alone anyone else's. Gus's 'murder' (let's allege this is what it was) offers a necessary springboard to consider the lessons we can learn from the wide spectrum of marine life that exist on our shorelines about how human life is impacted by decisions like these.

I took my son surfing in Kiama on the south coast of New South Wales and used it as my opportunity to snorkel at the famous Bushrangers Bay. I was hoping to see grey nurse sharks but that weekend they were absent. Ferocious looking but harmless to people, this docile shark has numerous ragged and overlapping teeth that jut out above the jawline. It looks like it's chewing on a bag of nails. They were almost annihilated in Australia by glory-seeking spearfishers in the mid 1900s and there are very few places left to see them today. They are extinct in Victoria. Meanwhile in New South Wales, some attempts to protect this critically endangered predator have failed as the powerful recreational fishing lobby doesn't want to be prohibited from fishing anywhere.

At sunrise my son and I took a stroll along a nearby clifftop. To get there you must navigate hundreds of hectares of new housing estate that almost touches the cliffside. Utes line every street and for every trade vehicle there are others parked in the driveway of new-build clifftop homes directly overlooking the sea.

Curiosity gets the better of me and before long I'm delving into the New South Wales coastal erosion exposure report. Sure enough, this is one of the areas at risk and there are many, many more.

Eighty-five per cent of Australians live at the coast. Forty percent of the world's population lives 100 km from the sea. Three-and-a-half-billion people worldwide depend on the ocean as a primary source of food, but only 15 per cent of the world's coastlines remain in their natural state.

What we know – and have known for a while – is that loss of fish life is the second greatest threat to coastline integrity. It seems Australia's love of both fishing and clifftop living aren't compatible. But what's the biggest threat? Man-made sea defence. That's right, engineered structures are considered one of the biggest causes of coastal erosion with 'varied and severe ecological impacts on coastal habitats'[51]. We, humanity, are building and fishing our way into an uncertain future.

The offshore reefs and seagrass beds that bind the sediment and buffer cliffs against persistent wave energy are falling apart because of overfishing. The fish that would diversify and sustain the resilience of these ecosystems are the larger megafauna such as wrasse that can live for decades and grow to tens of kilos in weight. Predators like the grey nurse shark are especially important because without sharks ecosystems collapse like a house of cards.

A study in the Bahamas has shown that the larger fish are driving patterns on forereefs[52] but these are the same fish that are targeted recreationally – especially by hunters and spear fishers. Fishing of this species and many others is putting entire coastal economies at risk.

LESSON 10: Value your local wildlife to save your lifestyles and livelihoods, homes and businesses.

Reinsurance company Swiss Re is responsible for insuring the insurers that insure you. From car to home, business to holiday they set the level of risk from which premiums are calculated. Swiss Re are warning governments and insurers that that the 're/insurance industry relies on functioning economies in which citizens and society can generate valuable assets and activities that are worth protecting' and is calling for 'nature-based insurance solutions'[53]. Swiss Re has identified Australia as the second-most-fragile country in the world when it comes to biodiversity and ecosystem services[54]. Coastal protection, erosion control and food provision are the top priorities.

Their report calls for 'Australia to prepare for ecologically driven disturbances – and look for opportunities in ecosystem services improvements and restoration' owing to its fragile nature. If we do not adopt laws that give nature the right to exist and uphold them, the consequence is that soon we may not be able to insure our properties or businesses[55].

In many places that's already happened as coastal erosion, fire and floods are making some homes uninsurable. Some of the most affluent suburbs from Sydney Harbour to the Northern Beaches are falling into the sea. Nature-based solutions can re-establish the integrity of offshore reefs fastest by restoring lost fish populations, but not if the oldest, largest and most important fish can be recklessly – even legally – killed for sport.

Ninety-seven per cent of wave energy can be dispersed by reefs[56] and those with more fish species do it much better [57-59]. Wildlife-driven processes might only be the last few per cent of this energy redistribution but this makes all the difference. If you're drowning, it isn't the first 1.5 m you sink that's the most worrying, it's the last few millimetres that cover your mouth and nose. This is what ecologists call a tipping point, and in Australia we have reached it.

When engineers talk about sea defence it tends to be focused on the more dramatic events like the '100-year storms'. But the biggest overall threat is the daily, low-energy gnawing away of our coasts. It is more insidious in that it doesn't threaten our daily lives but creeps up, drawing ever closer until a tipping point is reached. Then we have to rethink everything we thought we knew.

This is where natural systems of coastal protection trump anything made by man. Reefs, properly sited, don't exacerbate erosion elsewhere. When they're stocked with the right wildlife the creatures that inhabit them can adapt, modify and move to account for fluctuations and changes in the climate. Of the thousands of fish that inhabit the sea, only one or two species can do specific jobs. Hundreds of different species are needed for all the different roles to deliver a fully functioning reef. Lose one or two species and you can undermine the entire system.

Artificial defences create their own impacts that offset their benefit. They make things worse. Often this is because we can't build them on a big enough scale to work. For instance, we could stop cliffside erosion in Bayside, Melbourne by building a 10 m high stone wall, but the state government is unlikely to have the budget for a project of that scope. In Melbourne we're also fighting a battle against 'tree vandals' – otherwise known as coastal property owners who go out at night to clear the way for their ocean views. They certainly aren't going to agree to having their view blocked by a wall. Nor are the two million people who visit Bayside each year. Even if they did, flood defences are extraordinarily costly to maintain, which is money taxpayers cannot afford given that coastal erosion covers such a huge area. In the Caribbean, reef restoration has been shown to be significantly cheaper and always more cost effective than man-made breakwaters but given the chance nature does this for nothing and on a planetary scale.

In many communities spear fishing by locals to feed their village is part of nature. If it wasn't, the villages would have died out long ago. In Indonesia at villages like Sauwandarek in Raja Ampat, local communities have joined forces to ban all fishing except by villagers. In the seven years I've been visiting it's turned into an extraordinary spectacle of fish life where hundreds of reef fish congregate beneath the pier. Hunting for any means without a belief system, purpose and culture to look after sustainability has no connection to ecosystems.

One night in Naivivi I sat in a circle with some locals, drinking kava. Villagers built this resort but when they hear how much other people spend to stay in Fiji they laugh. The autonomy of the village is communal. The choices villagers make are not open ended, they are guided by a hierarchy based on cultural knowledge and taboo.

When US-based billionaires recently backed a scheme to reshape a natural coral reef in Fiji and create 'the perfect wave' for surfing it caused consternation. An adjacent surf resort has already sandbagged sections of reef, much to the horror of locals. The latest argument the villagers were given was that this reshaped reef would attract overseas tourists and bring more money for them. They unanimously rejected the billionaires' development citing one simple truth: the reef has always been there. It's there for a reason and it does not need to be changed.

This simple, concise statement spoke to the significance of nature in their lives. They prefer to protect their environment and culture not only because it defines who they are but because the reef is also one of their main sources of food.

This is the type of coherent wisdom we should all be using to protect ecosystems from reckless behaviour. The changes that are happening in our own cities aren't being led by a bunch of greenies. Instead, reaction to the death of fish like Gus the blue groper signifies a shift in public perception about nature. It is ordinary people like you and I who are beginning to realise the importance of our natural ecosystems and marine life.

As for the recreational fishing community? Well, without a common belief system in the importance of nature and strict enforcement of community interests, fish abundance declines. Species like blue groper have become a target for trophy hunters, and this represents a real and tangible threat to the lives and livelihoods of all people living on the coast, including the fishers themselves who are becoming tangled in their own nets.

The time is fast approaching when it will be commonly accepted that fish have far greater benefits to humanity when they are alive rather than dead. Fishing needs to be looked after by people who have the ecosystem's interests at heart for it to be sustainable. The fishing lobbyists need to wise up to this. It's not a simple emotional reaction and 'projection of human feelings onto the natural world', but rather it is exactly what needs to happen to protect society. There is resounding scientific evidence for this, and it is driven by stark economic reality.

Lessons from the Land

A petition arrived via email from a local lady wanting to save common bronzewing pigeons from the planned redevelopment of a local park. Bronzewings are pastel-coloured pigeons with metallic-looking wing feathers that reflect all the colours of the rainbow. These birds also call with a deep, resonating hum. I thought this was wonderful and signed straight away.

In the last 50 years, species loss has had 10 times more effect on our economies and livelihoods than all the major environmental threats we commonly associate with danger. Why does loss of another living entity have such a detrimental effect on our lives? Because it jeopardises the complex relationships between people and the ecosystems upon which we depend. The more connections there are between animals, the stronger nature's life support. Although only about half the world's species are declining, for our future to be secure we need most of these to recover and overall abundance to increase. The trouble is that once the connection between animals and vegetation is broken it can take tens of thousands of years to rebuild it back to full health.

Humanity's best hope is to act now. Restoring wildlife populations at this very moment gives us an advantage 10 times greater than any other action we can take to ensure our survival. Conservation is the most rewarding thing we can ever do and this is what rewilding programs are discovering.

Eliminating native species from land has caused the proliferation of invasive species into farmland. Near to protected areas there is less threat from acid soils and salination, plus increased benefits from pollinators and so forth[60]. In Europe, it's been calculated that a single herd of 170 European Bison introduced to the Carpathian Mountains is drawing down and storing carbon equivalent to 84,000 petrol cars[61]. The global potential for all wild animals is 6.4 billion tonnes of carbon capture a year, which would represent a 60 to 95 per cent increase in the productivity of our entire living landscape. However, when carbon is deposited through faeces, along with grazing and the brute force mechanics of a heavy bison creating ridges and furrows in the ground it increases the richness of plant species and engineers landscapes to restore soil integrity and water processes. The total value of this carbon capture is far greater than we realise.

Rewilding projects are more evidence for what we knew already: that nature can recover fast when given the opportunity. But scientists in Yellowstone National Park have found this hard to prove. By studying wolves using a conventional experimental approach, they have not been able to demonstrate the change they expected because the metrics are too linear. How can you know what to measure when you don't know how something will change or what signifies a good or bad outcome in the medium term? It turns out investigating the benefits of rewilding takes a leap of faith, while research takes more time than we have left to act.

Rather than let the hourglass run out while we wait for scientific experiments and processes, instead we can look to the animals that live around us and take these lessons from them. More and more people are acting by doing just this, having realised how important this is for our society's survival. It's fast becoming mainstream, seeping into national debates as leaders and key scientific advisors also start to believe in the importance of wildlife and its relevance to our economic and personal welfare in the future.

The late Thomas Berry, an advocate for rights of nature, argued that the abolition of slavery no longer relies so much on written laws as there is now almost universal acceptance of the human right to freedom by everyone on Earth. Berry believed that we could expect similar rights for nature to follow suit, and so do I. We need this unequivocal belief if we're to make wise decisions that ensure our survival.

Human beings live on this tiny sliver of land no more than a few metres deep. The soil beneath our feet is literally all we have left to stand on and that makes us vulnerable. Yet we are arrogantly dismissive of our fallibility as a species and instead consider ourselves dominant. You've probably heard figures in the media that farm animals occupy 30 times more biomass on Earth than all wild animals put together. It's quite amusing to think that if we keep going the way we are we'll create a planet where the pigs, cattle, sheep and chickens will become our successors.

George Orwell hit the nail on the head in his novel *Animal Farm*:

Man is the only creature that consumes without producing. He does not give milk, he does not lay eggs, he is too weak to pull the plough, he cannot run fast enough to catch rabbits. Yet he is lord of all the animals[62].

While the bronzewing pigeons that live in the park near me aren't the world's most iconic species and might only be significant on a local scale, they are still important. This is the crux of it.

A few decades ago, governments were investing heavily in development. We were all benefitting. It was humanity's heyday. We were healthy, mostly at peace, enjoying the countryside for holidays, booming businesses and the subsequent few decades would see the 'boomers' increase personal wealth like nothing before.

It was built on a bit of an illusion, though. Most of the investment was in 'nature-degrading' activities. Many of the benefits we were enjoying (including for our mental health) were being provided by nature. This includes those less tangible values that we can't necessarily put a dollar value on that make life worth living and that keep the climate for living stable so our businesses and lifestyles can thrive.

Wind forward to today and the full capacity for nature-based benefits has been mined, farmed, hunted, deforested and polluted. The naked truth is that it remains possible to squeeze out one dollar of value from nature-degrading investments to compensate communities for losses. However, that's not enough to rebuild damage done over the last few decades, and we are having to pay more and more money to compensate for loss of nature, risking further rises in inflation[63].

The good news is we can restore degraded ecosystems and regenerate between $7 and $30 in economic benefits for every dollar invested[41]. This is the whole basis for the European Union Nature Restoration Laws[64] that came into force in 2024.

Nevertheless, globally we are still spending 30 times more money on nature-degrading activities than we are on nature restoration[65]. Given that rebuilding ecosystems depends entirely on animal-driven processes, we all need to begin this journey to addressing our relationship with nature. That starts with preserving those species in our own backyard and working outward. Local, national, international and global – it is all linked by the natural world around us. Later in this book we will find out more about how these principles are being applied to wildlife in 'Lessons from Giants'.

I hope there will come a time when the local council would respond favourably to the petitioning local lady's request and give her a budget to employ an independent consultant who can co-design a plan to ensure the bird's ongoing existence. It is in this hope that we find a better future for humanity, because when it comes to rebalancing the Earth the value of even the humble wild pigeon is far greater than anything we could imagine.

Farms with Teeth

It was a balmy summer night and a light dew had set in. A halo appeared around the full moon overhead. To the south were the glittering lights of Melbourne, to the north a dark and featureless plain extended out to a horizon barely visible in the gloom. There was a familiar 'thump, thump' as a rock wallaby crossed our path. It paused nonchalantly at a granite boulder shaped like the upturned heel of a shoe, ruminated for a moment and then bounced away into the darkness.

When I first visited Mt Rothwell sanctuary in Melbourne's west in 2009 it was the only place left where the original mainland stock of eastern barred bandicoots could be found. Originally created by John Wamsley, this 420-hectare, predator-proofed reserve was bought in 2004 and operated by the Odonata Foundation. This private reserve has managed what no government-funded initiative ever could: it secured the bandicoots' future. Today, nine out of every 10 remaining brush-tailed rock-wallabies on Earth also have a home here and nowhere else. It's from these humble beginnings that animals like this will start to be seen elsewhere in Victoria in the future.

A visit to the Mt Rothwell sanctuary offers a glimpse into history; animals are everywhere. On that site there are estimated to be about 3,000 animals living in an area of 420 ha – about 714 animals per km^2. In comparison, Melbourne's human population density is among the highest in Australia at about 450 people per km^2 but across Victoria this is about 24 people per km^2.

We mostly live in an empty countryside when, in fact, our native fauna should be super abundant in these areas. More importantly we could be saving ourselves billions of dollars a year by recreating this lost abundance. And how do we do this? Simply by allowing wild animals to prosper in the landscape once more.

To see brush-tailed rock-wallabies, rufous bettongs, southern brown bandicoots and eastern barred bandicoots feeding together is normal. Over the past 15 years the conservation team at Mount Rothwell led by Annette Rypalski has experimented with recreating an entire working ecosystem that includes predators such as eagles and quolls.

On my first visit the sanctuary was newly acquired but the number of animals was already impressive. Returning 11 years later I could instantly see how the structure of the grassland had changed with their presence. There were more tussock grasses because the smaller animals had burrowed, dug and scraped their way across the landscape. By creating these micro-environments they engineer habitats for invertebrates, birds and other small mammals. In this place life begets life.

Sanctuary owner Nigel Sharp tells me that for the last few years, despite some of the heaviest rainfall on record, they've not had to do much fence reconstruction. The infrastructure has held firm. This is evidence that the soil is doing its job once again because the animals' burrows slow down and redirect excess water, leading to it being soaked up by an increasingly deep and fertile organic layer of earth. The sanctuary has also seen a decline in the need to manage weed pests. How? Because a quilt of tussocks and swales create vegetation diversity, reducing the opportunity for any plant – let alone a weed – to become widely established. The annual cost of ridding the land of weeds in Australia is thought to be up to six billion dollars. The evidence from the sanctuary proves that a gain in local wildlife translates into a huge economic and ecological benefit for society.

The benefits don't stop out there in the wilderness; encouraging nature is good for business, too. The declining productivity of land affects farmers most of all. As a result food production becomes harder, which has a knock-on effect on our economy and cost of living. But there are simple ways we can improve our farmland by doing things like preserving 'headlands' of habitat for wild birds and animals, as this has been shown to be effective at maintaining ecological processes such as soil and water.

Then Nigel founded Tiverton, setting up an impact fund with co-investor Harry Youngman and transferring this learning to a working farm. In fewer than three years they increased the wool quality and yield on an 800 ha sheep farm by reintroducing a species only just saved from extinction at Mt Rothwell – the eastern barred bandicoot.

Eastern barred bandicoots are small marsupials with a conical-shaped face and pink ears. They sport a waistcoat of striped fur over a hoary pelt and hop about the grasslands searching for grubs and insects. Once they were abundant, ranging across an area the size of Britain, but had been considered extinct in modern times. By chance a handful were rediscovered living among abandoned cars in a quarry near Hamilton in the 1980s.

Don't let anyone tell you we should give up on any species because it's too late or too hard. That's exactly what would have happened to the eastern barred bandicoot if individuals hadn't stepped in and acted to ensure otherwise. Twenty years back no-one imagined that this unassuming small predator could be unlocking new paths to restoring farmland profits. Planting more trees couldn't have achieved this outcome.

LESSON 11: Use predators not pesticides. They do it for free and without killing everything else.

By December 2023 the team behind the great bandicoot revival were rolling out a natural capital accounting project on dozens of farms to prove that restoring biodiversity and saving animals from extinction creates carbon capture benefits.

It has taken less than a generation to re-establish wild species and prove that can help reverse economic losses on farms. If we want a healthy environment to live in, we need to get the balance of animals right again. This means learning to live among them all.

Weeds are not the only pests that 'threaten' our farming businesses. After recent mouse plagues in Australia, the state and federal governments allowed the widespread use of second-generation rodenticides. However, everything we already know about farm pest management tells us this solution isn't even close to the right thing to do[66]. Small numbers of predators can have a huge impact on suppressing rodents, which is especially useful when rodents can erupt into plague numbers, but using poisons is putting these nature-based solutions at risk.

Rick Shine writes that 'snakes can collectively remove thousands of mice per km^2 of farmland each year and substantially increase farm productivity'[67]. He surmises that adult brown snakes could consume around about 10,000 mice per km^2, which seems to me a far more effective way of managing a plague of rodents than the risk of using poison that could get into the wrong animals' mouths.

In Florida, a dog kennel employee made a huge mistake by killing the snakes dwelling in the roof timbers. The subsequent rat plague ate huge amounts of dog food and took two years and hundreds of hours of manpower to control. Also, using second-generation rodenticide on whole landscapes not only endangers reptiles but also kills birds of prey who feed on any other animal that has consumed the poison.

My first thought after reading Rick's article was that reducing snake numbers around grain sheds could create a 'super-spreader' event because, as we all know, mice breed very quickly. When conditions are right a small surplus can create runaway reproduction cycles.

Meanwhile snakes – like mice and birds of prey – have become common enemies of the modern-day farming system. Mostly the vendetta imposed on these predators in our farmland areas is counter-effective and unnecessary. Snakes are largely shy and harmless unless provoked but are still attacked on sight in many cases – usually with a shovel. Even though living with wildlife is clearly a more sustainable outcome, this refusal to understand and believe in the importance of nature only worsens our current problems.

Being the curious chap I am, I dig deeper into the subject and find a paper titled, 'Spatial variability in ecosystem services: simple rules for predator-mediated pest suppression'[68]. 'Simple', apparently. Well, simple for anyone who already believes in the importance of animals.

The scientists built a model of 1,000 simulated landscapes with two types of predators:

- poor dispersal (for example, snakes or other animals with limited mobility)
- good dispersal (for example, birds of prey that can fly longer distances)

They then varied the distance between patches of predators and patches of prey. Unsurprisingly, they found that more mobile and far-ranging predators provided the best pest suppression in most landscapes but for one exception; for poorly dispersing predators, patches needed to be especially closer together.

Overall, farms with a balance of predator and prey living close together have lower risk of pests. We didn't really need all that science to tell us that, it's just common sense. You need snakes around your grain sheds where the mice occur at especially high density and owls, kestrels and other birds of prey ranging safely across the whole landscape to maintain overall balance.

Farmers who kill even a few snakes are compromising their livelihoods, but anyone who uses rodenticide is poisoning the entire food chain. Flagrant misuse of pesticides risks creating another 'silent spring' by wiping out every bird of prey essential for our livelihoods and food across whole landscapes. Second-generation rodenticides are so toxic that consumption of one poisoned mouse is enough to kill any bird outright.

This is not a Band-Aid solution, rather it is akin to thinking something like the tip of your finger hurts, so you cut it off. Then thinking the rest of that finger hurts and cutting that off. Then your hand is hurting, so you cut that off. Pesticides are an unnecessary and rather cynical amputation of an essential limb when all we need are Band-Aid solutions – that is to say very carefully and precisely administered temporary measures designed to keep problems at bay while buying nature time to restore itself. This is about being 'nature-positive' – in other words, protecting the underlying system that maintains the life support we depend on. Poisons do the opposite, yet the abuse of them has become doctrine among our conservation and farming leadership. Nature 'prunes' populations like a bonsai master but this is something we can't do ourselves. It's not practical. Poison is a blunt tool, and military-style warfare on wildlife isn't going to be of any help.

When Rob Farnes and Les Thyer were both living in Mackay in the 2000s they used to visit a grass owl site at Kinchant Dam. 'One night we counted 82 glowing eyes as we scanned the grassland with a spotlight,' Rob recalls. Then there was a rat plague in the sugar cane fields and the state authorised use of a potent rodenticide. 'The farmers were mixing far too much of it,' he says. 'We went back to the site later and found dead and dying owls on the ground. It was awful.' The move probably wiped out 90 per cent of the birds in that area.

The belief systems of those given custodianship of our ecosystems aren't yet compatible with the needs of society. These overseers aren't yet thinking about the consequences but they really need to. Given the clear and compelling risks to our entire civilisation, using pesticides could be tantamount to ecocide. Ecocide, which has a legal definition, means 'unlawful or wanton acts committed with knowledge that there is a substantial likelihood of severe and either widespread or long-term damage to the environment being caused by those acts'[69]. Proposals to include ecocide in legislation are starting to progress through parliaments worldwide.

Society currently lacks belief in the overriding evidence that nature-based solutions are by far the best investment in farm efficiency. There is still an unfortunate schism between those who know this and the scientists who advise policymakers and leaders, whose personal beliefs fall short of making the connection between predators, landscapes and ecosystem life support services.

There is only one fact we know for sure. Landscapes 'with teeth' that retain predators throughout the food chain are far more resilient. Animal plagues can never be stopped completely, even by sparingly using chemicals (which is all we can afford to do if we don't want to undermine everything else). Nonetheless, it is a fact that the frequency of occurrence and intensity of these pests will be massively reduced by rebuilding ecosystem structure and function. In the case of mouse plagues, this means ensuring the abundance of predators like snakes and birds of prey[70].

Inside the Mt Rothwell sanctuary and at Tiverton, animals run wild without the overuse of pesticides. Tangible benefits are had within a matter of a few years as wildlife goes about its business eating each other.

Now there is a waiting list of farms eager to adopt these nature-based techniques. All over the world collapsing farm economies are trying their hand at using nature to reverse losses. The results won't be immediate, they can't be because nature doesn't work like that. Instead, it's messy and operates more like a building site – it takes time for constructions to take proper shape.

One thing is for certain, though; once there is a belief in the power of nature to restore, the use of poisons and chemicals becomes moot. As the late Thomas Berry said, then there will be a common acceptance that nature has a right to exist as we will learn to give animals the autonomy they need without jeopardising our best interests.

Aerial Architects

During lockdowns, 80 conservation representatives and policymakers sat on Zoom meetings, disconnected from the natural world we love, discussing how to prioritise spending on threatened species in a one-million-hectare region of New South Wales. The plan was to map threats and actions for a dozen species or groups of species occurring across this vast expanse of our landscape. By the end of it all there was one outstanding result: just eight actions applied to these dozen species would address 95 per cent of threats to them and their habitat. It was heartening to know that everyone was confident this could be done.

At the end I asked everyone what benefit restoring these threatened wildlife populations would bring to people. 'Tourism' was a dominant response, even though among the most important findings was the need to recreate permanent standing water through the dry season. Protecting water resources is essential to farm productivity, human health and the economy, yet most answers failed to make this connection.

Australia's leading fruit bat ecologist, Peggy Eby, was the only person to mention the fact that 'fruit bats generate their own ecosystems'. The simple wisdom in Peggy's words is a fact that eludes many ecologists; the kings of forest ecology are its animals. As well as being the keepers of their own habitat, animals are also the builders and custodians of ours. As Peggy observes, our entire life support depends on these winged wonders.

Australia's grey-headed flying foxes (another name for fruit bat) are nomadic colonialists. In a single year they transit through many dozens of roosts on a constant quest for food. Peggy prefers to use the word 'camp' rather than 'roost' as these residences are more like motels than a permanent home. Each night the bats fly out and disperse over distances of 50 km or more, and over a single season it's not unusual for fruit bats to travel 6,000 km. One particularly intrepid individual clocked up 12,000 km in total between Melbourne and Brisbane.

A billion years ago the Earth was clothed in ancient vegetation and humans could never have existed. Today we inhabit a world where ancient trees are part of the landscape that animals made for themselves. From the jay that stores acorns, to the passenger pigeons that used to strip oaks bare, animals have created the structure that turns habitat into ecosystems for us all.

There are few species better equipped than fruit bats to help rebuild a habitable world. The good news is that a large flying animal like this can cross farmland, roads and deforestation corridors – meaning the benefits of their constructive nature reach far and wide. When it comes to building ecosystems, colonial migratory animals are like an amorphous being that threads its way across entire nations, continuously maintaining ecosystem processes.

Bats move in a coordinated way, sharing knowledge about where to find food and transferring nutrition back and forth to create centralised resources that fertilise the land. Of course, they also spread seeds and help forests germinate.

There are estimated to be up to a million grey-headed flying foxes in Australia, which means collectively they could be travelling over 5 billion km annually – the equivalent of circumnavigating the Earth 110,000 times a year.

Every individual is making intelligent choices on where to feed based on knowledge passed down over thousands of generations, adapted year to year by communicating with fellow bats so the overall patterns of consumption and creation of resources become balanced with their own needs and that of all other animals.

As humans value timber, healthy forests and climate, that is a clear indication that we should also care about fruit bats. Until recently, however, our city's bats were under constant threat from councils and local people who hate them with a vengeance. Despite urban spread and constant redevelopment, they somehow manage to maintain their ability to find food. They have even mapped where 100-year-old trees grow among the Melbourne skyscrapers. At their roost at sunset, the air is filled by the scent and sound of a million bats defecating, socialising and sharing knowledge.

LESSON 12: Animals that fly reward us by spreading their impact far and wide, creating inestimable benefit for humankind.

Bats are not the only invisible builders in our world. Hornbills serve a similar landscape-building function over the fortressed forests that are surrounded by steep limestone escarpments in West Papua. Against a purple and orange sunset they arrive in social groups of twos, threes and fours, their deep wingbeats catching the humid air with a whoosh. We might think they gather for security – that is part of it as these birds may be hunted – but for the most part it's a fruiting-tree information exchange.

One morning we head upriver and deep into the mountains to watch the world emerge from slumber. You can easily believe these forests are empty if you go during the warmest part of the day when everything is quiet, but go an hour before dawn and you soon realise every inch is occupied and bursting with life. The treetops are still clouded in mist when the chorus begins; jewel-babblers are the first with their falsetto whistle penetrating the cool air. As the humidity begins to lift and the air warms, other birds chime in with different vocal instruments. The last are the tubas of the forest, the deep-throated pigeons whose sounds provide a resonant bass to this orchestral performance.

Just before sunrise the hornbills find their voice. These enormous hammer-headed birds with scythe-like bills, skinny necks, broad shoulders and long tails assemble at the top of a dead tree. Two of them duet to each other, the sound akin to two people hurriedly using hacksaws to cut through metal. This questionable 'singing' creates bonds between birds of knowledge-sharing that they depend on to find food. But those secrets aren't shared with the whole group. Like teenagers at a dance party they gather in cliques, the coolest birds assembling with like-minded members at the exclusion of others. Fruiting figs (their favourite food) are a scarce resource, which means the integrity of the entire ecosystem (as with bats) is dependent upon a minimum number of birds knowing where to go. The forest needs these secret keepers who are more adventurous than others, who can reach further and find new feeding grounds and who can adapt their culture to survive in an ever-changing ecology. Their survival begets the forest, which creates a habitable Earth for us too.

Fruit bats have a similar purpose but unfortunately in Australia their numbers have declined and their forest habitats are increasingly fragmented. They have drifted from being comprised of a small number of large colonies to a larger number of small colonies. This dilution is symptomatic of a degraded ecosystem where food is less available and harder to find. There's less to feed a large colony, so hanging out with fewer friends makes them more agile, adaptable and less likely to starve.

Bat civilisation and culture are changing; they are creating their own new world order, not because they want to but because they have no purpose without. Their survival as forest engineers is what their species depends on.

Humans co-depend on the bats' remarkable ability to change their behaviour and start the laborious process of rebuilding a habitable ecosystem for all animal kind. We can't continue to survive in a degraded landscape, which means our own future now depends more on the bats than on any actions we take alone.

Ghost Builders of Borneo

Orangutans have a quiet demeanour and slow pace of life that perhaps reflects their need to carry on with their daily routines of forest engineering. Every one of the 11,000 nests an orangutan creates in a lifetime is about a metre in diameter. To build them, the animal breaks branches and weaves them together then furnishes it with leaves. Nest-building is a chore as a new one must be constructed every night.

Scientists have observed that orangutan nests tend to be associated with taller trees and more uniform canopy structure. This led to an article by the Orangutan Conservancy called, 'How canopy structure affects orangutan nesting sites'. In fact, it should really have been called, 'How orangutan nesting creates canopy structure'[71].

Building nests is a function of orangutan ecology. Most primates (including humans) prepare their bed before nightfall, but for orangutans bed-making is also an act of ecosystem engineering because shelter and survival are entwined in millions of years of connection to their forest home. It's common to see these structures dotted along riverbanks where these creatures like to live.

Ian Redmond, who worked as the late primatologist Dian Fossey's research assistant, added up the number of these structures throughout an orangutan's life for the magazine *Primate Eye* in October 2021.

> Each adult orangutan builds a new nest every evening, and by constructing it, creates a light gap in the canopy – 365 days per year. Nest construction is a bit like folding an umbrella – it makes a light gap by bringing leaves at the tip of each branch used into a tight ball of vegetation, which forms the sleeping platform, like a giant bird's nest. An orangutan might live for 40 to 50 years, so if we take 40 years as a reasonable age, he or she is likely to be making nests for 32 years or more. This equates to more than 11,000 such gaps in the canopy per orangutan[72].

It is wrong to assume that it's the trees that build this canopy structure. Orangutans have been around for millions of years and for all that time they have been creating light gaps. Sunlight is the harbinger of all new life, and by making a complex canopy structure the orangutans diversified the forest so it could produce stronger trees, more fruit and support a vibrant mix of other wildlife.

In the 1980s there were an estimated 21,000 orangutans in forest reserves and state parks in Malaysian Borneo, an area of about 29,000 km^2 [73]. Males can even have territories of 100 km^2 but are more common in lowland forest below about 500 m. In prehistoric times orangutans would have been even more abundant and widespread[74] but their population contracted after the appearance of another primate species – humans.

People became a substitute for orangutans and indigenous humans like the Dayak in Borneo were (and still are) part of ecology. Like hornbills and people, orangutans are social animals with a propensity to share knowledge. Beneath every nest are piles of seeds and manure that fertilise the newly sunlit ground, mobilising soil microbes and nurturing fungi, plants and insects. The unimaginable intensity of orangutans' actions, their vast distribution, comprehensive knowledge and tree-climbing ability means there is hardly a patch of their tropical rainforest home that isn't built by them.

Here I could list the many reasons we know how they do this such as fruit dispersal, defecation and so on, but truth be told in the complex interrelationship they have with the forest most connections between these builders and their benefits are unknown.

If these animals had not contributed to the forest's wellness by gathering food and building shelter they could never have lasted for millions of years, as every action they took would have contributed to their own collapse. This is a key point that humans need to learn: survival is about both give and take. Our own actions for harvesting food, water and building homes take and animals help us give back.

Living among the orangutans of Borneo is another ghost builder, the sun bear. These animals are rarely observed and even when humans catch one it offers no real explanation for its existence. And nor should it have to. In fact, the forest could not exist without them.

Sun bears are the smallest bear and weigh about as much as a human. They are rarely seen in their natural habitat and because of this they are one of the least-studied bear species in the world. The fact we don't know much about them doesn't make them any less important. This lack of knowledge means the work of Dr Wong and everyone at the Bornean Sun Bear Conservation Centre is vital. The research conducted here will help secure the future of the forest and the life support it provides, alongside showing the world that sun bears have an incredibly important role to play in our ecosystems and must have the right to roam wild like every other animal.

Sun bears are essential to the health of the rainforest. They are forest planters eating fruit and dispersing seeds throughout their habitat that keeps the ecosystem healthy. They are forest doctors feeding on termites that threaten to fell trees if they are not kept in check. They are forest farmers tilling the soil with their strong claws as they dig for food and enhance the nutrient cycle. And they are forest engineers shaping homes for rainforest creatures[75].

Over the course of a year, sun bears might walk 550 km – the equivalent of half the width of Borneo. They can forage for 20 hours a day at certain times of the year[76]. During that time sun bears will break branches, build nests and raid beehives, shaping the rainforest canopy, mid-storey levels and forest floor in the process. They will eat all the most abundant fruit as well as termites, ants, bee honey, larvae, beetles, earthworms, bird eggs, reptiles, small animals, mushrooms, succulents and flowers.

Fruit is their main diet when they can get it and they are particularly fond of figs. Sun bears have been found to consume over 100 types of fruit and use nearly 800 species of rainforest tree to nourish themselves. How much they eat, however, isn't as important as where, how often and at what intensity.

Black bears can smell food over a mile away and this also holds for sun bears, too. If this is the case then each bear is sampling from a menu that stretches 1,650 km^2 each year. This means bears home in on and consume the most important food sources 110 times a year over an average territory size of 15 km^2. Compare this with the fact golf courses routinely employ 10 to 15 ground staff over an area of only one km^2 and we see how effectively animals like this maintain our landscapes. It's clear that if we want to look after forests, we're better off giving animals their autonomy – it's a lot more efficient than human efforts can ever be.

> **LESSON 13**: Just because an animal seems invisible to us doesn't mean it's not making an important contribution.

Because animals are virtually invisible to us, trying to fully understand how they exist is a feat of impossibility. There are a million stories behind every one, too many to comprehend in fact. Unfortunately, this also means we can never hope to study and prove the connections they have to our lives. We must take a leap of faith and accept their importance if we're going to survive.

While the livelihoods of animals remain invisible to us their behaviour appears alien and the reasons why they matter will rarely emerge from our subconscious, but this doesn't negate the fact that they are vital to the survival of both our species and the ecosystems we live within. The same goes for humans. If you are behaving as well as you can, your inaction and sobriety means you are living more in balance with the world – a trait essential to our species' survival and humanity's future.

For the most part, the animals we rely on are some of the largest of their kind despite the fact most are unseen – invisible builders of our world whose behaviour is a shadow of ancestral haunts that have lasted literally millions of years. Their culture is engrained within forests and they are as an integral to the trees as leaves are to branches. So, perhaps, when you next watch a wildlife documentary or, if you're fortunate, stare out over the canopy of any rainforest you can now rethink your view of that structure. This landscape was made by animals, and they are still there working away, engineering the world for us even if we cannot see them.

Living on Burrowed Time

Many animals such as migratory birds and wildebeest are temporarily present in an ecosystem, while others like insects and rodents are tiny, fastidious permanent fixtures in certain habitats. But aardvarks are special because these slightly odd-looking creatures are some of our most ancient residents.

Fossil records indicate the presence of aardvarks as far back as 5.4 million years ago. That's 5.2 million years before hominins evolved and more than twice as long as any human-like animal has even existed. Now, we all know that age doesn't always necessarily mean wisdom (despite what some people might tell you) but in the case of aardvarks it absolutely does. Their behaviour and physical impact on the land has been shaping the desert for longer than it's possible for the human brain to imagine.

Aardvark burrows are enormous and the density of these dwellings in a single territory is quite extraordinary. They use them for several days at a time then either dig a new one or renovate an existing hole. As well as giving shelter and nesting sites for a wide range of different animals, these tunnel networks also alter humidity patterns by providing places for water to drain and disperse underground. They also create a new subterranean ecosystem for a great diversity of life to thrive in.

To understand the significance of aardvarks we must first consider the concept of size in relation to both humanity and the ecosystems we exist in. As we already know, the main thing that sets animals apart from plants is their ability to move. Our floating brains allow us to turn knowledge about the Earth into autonomy, to find and manipulate the resources around us and create a habitat in which we can thrive. Though migration is the ultimate tool that enables bulk transport of energy, the world also needs animals that are permanent residents to keep systems working during down periods and continuously construct and maintain key services. A live-in groundskeeper if you will.

When we think about ecosystems we must also think in terms of different scales. It might surprise you to know there are ecosystem engineering genes inside your body – chaotic genetic material that can adapt to create different proteins and function in a changing day-to-day environment. For example, exercise changes the way our DNA interacts with some of these, triggering the production of proteins that build muscle.

Ecosystems are like Russian dolls – each one contains a smaller subset of an ecosystem that works in just the same way right down to the molecular level. But when scientists talk about animals as 'ecosystem engineers' they are usually referring to the largest creatures. This is partly because humanity is biased towards believing animals that are as big as us or on a scale resembling our species' world view are more significant. It is important to understand this is a fallacy. Humans didn't evolve into a world where we were just slightly bigger than other creatures that came before, instead we exist about halfway up the food chain surrounded by animals both smaller and bigger us. The bigger animals we need because they create ecosystem disruption and concentrate resources in large enough quantities so we have what we need to survive: food, water and shelter. We also used to be hunters, which means we have quite large brains (although they have shrunk somewhat since we stopped hunting huge, mobile animals like mammoths and begun farming).

Large animals create the foundations for the patterns in those smaller-scale boxes to evolve and through that process life can thrive at the tiers we need to create our food, water and climate. Without animals such as aardvarks many of the other species – humans included – will find it hard to make a living in future. Bigger animals (or abundances of colonial animals) create larger patches of nutrients that enable large-bodied counterparts to coexist. Human physiology changes depending on our access to resources that animals help construct and maintain; as we live roughly halfway up the food chain, we depend more than ever on being surrounded by animals similar in size to us.

Tyson Yunkaporta, a Deakin University scholar and the author of *Sand Talk*[77], wrote on LinkedIn in October 2023:

> Last time we had this kind of climate change (drying and warming) our megafauna scaled down to human/roo size. This time we think smaller, like rat/cat size, and that's what we'll be hunting for meat if we can't afford a $500 beefsteak. Over two centuries we humans would shrink as well.

Aardvarks are big. They weigh up to about 60 kg and can reach two metres long. It's not just their size that matters but also what they do, how they do it, how intensively, how often and for how long. Their impact has been forever and it doesn't have to be that intensive because they have built deeper foundations than almost anything else that exists around them.

This prehistory makes aardvarks just as integral to the landscape of today as a rock, sand dune or mountain. This fact is rarely acknowledged in conservation science where animals are seen as an addition to the landscape rather than part of it. Our science tends to look at current impacts in terms of a timescale of decades or estimates possible futures on a scale of years. It has only taken a few hundred years for people to wear down the marble steps on Italy's Tower of Pisa, so what do you imagine was the effect of aardvarks walking circuits around the landscape and digging the soil over millions of years?

It's more common to read older historical truths from indigenous science, which is the only human history we have to guide us. So often we disregard this knowledge in favour of our short-sighted modern science, but as Tyson says:

> There is Law and knowledge of Law in stones. All law-breaking comes from that first evil thought, that original sin of placing yourself above the land or above other people[77].

Instead of placing ourselves as rulers of the land we need to bow down to the wisdom of those – human or not – that have come before us.

> **LESSON 14:** Animals have been creating a habitable planet for longer than we can possibly imagine.

As we break ecosystem processes by altering weather, soil and deforesting landscapes, we fragment these million-year-old connections built by animals that existed long before us. It is the loss of wildlife that destroys ecosystem stability, not the loss of habitat. We can plant trees and habitats can regrow, but unless animals remain adequately linked to the landscape they helped to make then we lose the patterns of stability for food, water and climate that are everything for our survival. If animal science is to become relevant to the future of humanity, biologists should be asking different questions to the ones they are now – questions framed by an understanding of the role of wildlife in the processes and functions of ecosystems on Earth.

When you look at a desert you might see sand, bushes and grass. An ecologist will see a mosaic of sweeping and discernible patterns in the undulations of geology and dunes – places where moisture can become trapped. However, landscapes are not ecosystems unless they contain wildlife and it is animals that fill these damp depressions with life-giving nutrients that transfer, amplify and concentrate resources. Piled into places where and when other animals can expect to find them, they offer certainty for survival.

Aardvarks are the perfect example of this because they are walking containers of nutrients. They feed predominantly on ants and termites and then defecate the waste into grassland. This happens in patches that other animals discover and exploit for their own sustenance, storing this knowledge in their floating brains to ensure their survival.

The most significant behaviour of aardvarks is burrowing and just like humans their need to find shelter to sleep forces them to repeatedly return to the same places. The movement of aardvarks in a landscape isn't random, it's based on centuries or perhaps thousands of years of work. Burrows and foraging digs are a predominant feature of the African savannah and this has led to aardvarks being called 'engineers' of this vast landscape. As you can imagine, the burrows are massive!

Research shows that aardvarks tend to use a single burrow for about five to nine days before either renovating a previously made one or creating a new one. Stephanie Ann Martin at Nelson Mandela Metropolitan University in South Africa collected data from just under 129 burrows over 40 ha of land in deserts of the eastern Karoo[78]. However, when her research included foraging holes the number of digs increased to nearly 2,000 in that small sample of the landscape.

The scale of impact that aardvarks have on the landscape is significant, especially when combined with the magnitude of their effect. They continuously dig soil for foraging and burrowing. In aardvark-ideal altitudes and habitat about 140 holes have been counted and the amount of soil turned over just for foraging can be almost six tonnes per ha. This means the average foraging aardvark might be disturbing 180 tonnes of soil at the surface each year. Add in refuge burrows and this increases to several times that volume.

How does that look in real-life land terms? Let's take southern Africa as an example. Distribution of aardvarks there is patchy. Where they do occur, such as in the Kalahari, there might be perhaps eight animals for every 10 km² and suitable habitat probably covers one-third of the landscape. You'll find them mostly in flat or gently sloping areas of semi-arid desert and savannah, which are the predominant landscape types throughout Sub-Saharan Africa. If there are populations of 10,000 or more animals this would equate to an area of about 12,500 km² or about 3.3 per cent of suitable habitat. A few per cent might not seem significant but in percentage terms it's equivalent to half of all the arable food production land in South Africa.

In short, aardvarks are by far the biggest soil engineers in the ecosystems they exist in. However, there is another character in the aardvark tale as they don't engineer this habitat alone. Other animals like porcupines, bat-eared foxes, meerkats and so on, can't or don't always dig their own burrows, preferring to use holes made by aardvarks. The rare blue swallow nests inside aardvark burrows, bringing thousands of years of phosphate-rich guano – the type of fertiliser that launched global agriculture. Aardvarks create the structural, moisture and nutrient conditions for ants and termites to thrive. They promote vegetation growth that insects break down with the help of fungus and disseminate back into the soil alongside microbes. Aardvarks' intensity of impact on these landscapes is amplified by their connection to these other animals.

Permanence is another significant factor in the balance of this ecosystem. Given that animals must return to burrows each night, their impact intensifies over preferred areas and creates the patchiness needed for smaller creatures to find reliable food. Though aardvarks have surprisingly small home ranges and only forage in the space of about three km², within that area they can travel up to 30 km each night. Now, when I say they hunt all night I mean they hunt *all* night and by continuously turning over the soil layer they affect hundreds of square metres of habitat. This is something they've been doing for millions of years and humans and other animals in their ecosystem have reaped the rewards.

You'd be wrong in thinking this perpetual farming is specific to aardvarks. In fact, take any large and reasonably abundant animal and you'll find hardly a patch of ground untouched. Though the impact of their actions might be invisible to us, it is still significant.

To assess the conservation status of animals using only distribution maps and numbers underestimates their importance in a landscape. In global conservation listings aardvarks come in under 'least concern' because they are widespread. Though they are naturally low density, the listing doesn't consider the intensity of their effect on the landscapes where they do occur. That they are vulnerable to extinction in all settled areas and extinct in many places with a high concentration of people is not the most pressing point. Rather, it's the lack of understanding of how critical they are for the ecological and economic integrity of entire nations.

Don't imagine this story is unique to Africa, either. Wombats in Australia have just the same effect and they also provide refuges for other animals to escape bushfires. Badgers and foxes burrow in the UK and in the US so do coyotes, groundhogs and gopher tortoises. These subterranean constructions are the mark of our unseen architects. Without abundant burrowing megafauna our landscape loses what we need to survive.

These creatures have built the world for us and it is up to us to ensure that they do not go extinct. When we allow that, even locally, we permanently erase the knowledge of creatures that had evolved to retain the balance of the Earth. We break the thread of their ancestral culture that they communicated to us through their actions and behaviour and lose something vital to our own survival.

Humans tend to think in terms of years, decades or lifetimes, but landscapes are set in stone by animals over eons. The results from wildlife recovery are nothing short of miraculous when given chance. They have sculpted the terrain for longer than the human brain can comprehend and if we allow them can continue to create the foundations for a global system of nature that looks after our food and water.

The Surprising Elephant Economy

We were crossing the border from India to Nepal though a tiny town called Nepalgunj. It had taken a week to get there on local buses. Each day we'd pass through balmy Himalayan foothills where our lungs were filled with clean mountain air and the roads were lined with blossoming trees and lush green rice paddies. Distant snow-capped hills provided the backdrop.

We were headed to meet Balbahadur Ranamagar, or 'B' for short, who owns a small lodge in Bardiya National Park where we planned to spend a couple of weeks. As we crossed the border we found our bus, as instructed. Gaggles of kids were walking to school, their perfectly neat blue-and-yellow-trimmed attire contrasting with the dirt-lined streets. A few clambered aboard with us, settling onto sparsely cushioned plywood seats and stealing the occasional stare, giggling at the strange sight of foreigners so far from their normal habitat. The driver adjusted the Bollywood music to a level just above comfortable before we wove our way west in between cattle-drawn carts, tuk-tuks and ornately decorated trucks.

B met us at the entrance to the park and we headed to his forest home on the banks of the Girwa River. Our main intention was to see wildlife on foot as Bardiya is home to Bengal tigers, one-horned rhinos, Asiatic elephants and other exotic fauna such as great hornbills, grey langur, Ganges river dolphin, floricans, blackbuck and nilgai.

You always want to smell a rhino before you see one, despite the fact they have an aroma something like a cattle yard. Why? Because when you smell them it means the wind is blowing towards you and they probably can't see you. But also because (being seriously short-sighted) rhinos will charge when they smell danger. Avoiding them is easier said than done, though, as you march cautiously through 10-foot-tall elephant grass past mudholes where the rhinos like to wallow. The first rhino we saw was downwind, thankfully, but for safety we climbed the nearest acacia as the rhino snorted, proffered its horn in our direction with forward thrusts of its wrestler neck and stamped the ground with thundering hooves.

Nepal has been immensely successful in rhino conservation; from a historic low of about 100 animals in the mid 20th century, there may be over 800 today. They are even using them to rewild reserves where their digging, dung and latrines fertilise landscapes, restore water courses, disperse trees and provide focus for other animals to congregate and amplify ecosystem processes.

Our several hours our walk went deeper into 'man-eater' terrain in search of a tiger. As we encountered a bushfire ahead and had to turn back – right through the territory of a female tiger recently responsible for killing a local villager – a herd of nilgai fled the flames, coursing through the Terai ahead of us. Back through the jungle we went, heading to the river for a packed lunch whereupon a lone, wild, bull elephant crossed in the shallows.

If living with tigers and rhinos is hard, living with elephants is practically impossible. Elephants are highly intelligent herd animals with a belligerent attitude, a propensity to want to eat exactly what we grow and erratic mood swings. But the process of increasing tiger populations in West Nepal has meant conserving forests and creating vegetated movement corridors, which has led to marked growth in the local elephant population.

For even the most tolerant of societies, increasing urbanisation and declining habitat area inevitably results in escalating human–wildlife conflict. Nerves become frayed and before long both elephants and people become agitated. In India, conservation scientist Krithi Karanth launched Wild Seve in 2015, which now provides over 2,000 villages with compensation for damage to their land. Over 25,000 claims have been filed with the majority being for crop loss. In 94 per cent of cases elephants are the problem, but more on that a little later.

Understandably people can be reluctant to support measures to increase elephant numbers as they are naturally scared by these creatures, especially when moves are backed by powerful conservation lobbyists comprising global organisations connected to wealthy and influential politicians. In the case of many indigenous groups, the legal system has already fractured their way of life given that all over the world the creation of national parks was done at the expense of forest-dwelling nations. For instance, in the last few decades over 100,000 such people have been displaced from their homes in India to make way for parks to protect 3,000 tigers. The Kattunayakan Tribe that featured in the striking Oscar-winning documentary *The Elephant Whisperers* practised belief systems that viewed animals, trees and even rocks as sacred[79]. The film showcases the ancestral tradition of caring for elephants, but the Kattunayakan were also hunter-gatherers whose freedoms were removed by the creation of national parks.

To a large degree catering to tourism has replaced traditional ways of life and while that has filled the void with much needed income and supported conservation, it won't last forever. Tourism is a fickle market prone to fits and starts, forced by complex interactions between domestic and overseas cost of living, exchange rates and media trends. During COVID, WWF Nepal reported an uptick in poaching of tigers and rhinos, blaming the collapse of ecotourism and lack of income. But tourism is an easy option; in the long term we need to nurture and regrow traditional belief systems and support the remaining few custodians of natural wisdom as they look after the animals on our planet as part of their coexistence with nature.

Working with communities is the best way to do this. Conservationist Ignacio Jiménez, who has coordinated the largest reintroduction programs in the Americas, says:

> The most effective programs connect to the needs of most people. Once we reframed the whole idea with restoring culture, restoring nature and providing economic opportunities to forgotten rural communities, they were much more interested in it.

What if there was a way to look after these stewards of nature, to help them to protect wildlife in perpetuity and even boost its abundance while ecosystems reform?

Rebalance Earth was co-founded by then International Monetary Fund Assistant Director Ralph Chami, financial activist Robert Gardner and technology strategist Walid Al Saqqaf to focus on nature as an investable asset. Their scientists may have discovered a way to transition to a new nature-based economy that doesn't depend on tourism. How? It all begins with carbon.

It turns out the great forests of the Congo Basin store more carbon than the Amazon River basin – a surprising discovery that contradicts popular scientific opinion. Why? Simply because there are more elephants in Africa than in the Amazon and their action (and the reaction of other creatures, including people) traditionally returns more carbon to the ground. Once upon a time the Amazon did have elephants – rather, it had large elephant-like herbivores. They were called *Gomphotheres* and the last of these died out about 9,000 years ago.

The team at Rebalance Earth originally set about calculating the value of a single animal based on the amount of carbon it created over a lifetime, starting with whales (we'll find out what happened next in 'Guardians of the Wild'). They weren't trying to include all variables, just the part that describes the amount of carbon an animal itself sequesters until it dies. Using a conventional economic valuation, they treated carbon like a share of stock where it continuously gives rise to more and more dividends over an animal's lifetime. They then added up the dividends, multiplied these by the value of a tonne of carbon and discounted this all the way back to the present. This gave them a minimum price on a *living* whale. They quite literally put a price on wildlife – but for more positive, posterity-led reasons than hunting could ever bring.

The carbon value of elephants is huge. Letting elephants roam free to have their autonomy is worth upwards of US$2.3 million of carbon sequestration per elephant. Compare this to the value of selling an elephant's ivory – about US$40,000 – and it becomes clear that our wildlife is far more valuable to us alive than dead.

Rebalance Earth and CEO of Blue Green Future Ralph Chami explains in his TED Talk that the country of Gabon in Africa has 57,000 elephants. If we left them alone, that number could rise to 195,000. The potential value of Gabon as a country when considered in the currency of living, undisturbed elephants is a minimum of US$30 billion a year. This is money that Gabon could store as an asset in its country's capital accounting and use it to attract nature-based funding.

The Rebalance Earth team may have just rewritten our market definition of 'nature'. For me, this discovery is as significant as when Dr Jane Goodall found the first example of tool use in animals, leading to the rewriting of the definition of 'man'. Or when Biruté Galdikas first communed with a male orangutan. Why is this finding by Rebalance Earth so significant? Again it shows that our animals are far more valuable to us alive than they ever will be dead.

There is a rocky road ahead for the elephants of Gabon, though. A change in government in the country has renewed pushes to exterminate elephants. That government is sceptical because Western governments have long reneged on promises to remunerate Gabon for conservation and instead built them debt. Rebuilding trust isn't just about restoring respect between species but also between our own cultures. The damage from colonial rule and years of oppressive lending policies makes unwinding the opportunity for a nature-based economy more complex than it ought to be.

My question is, why have we not thought to consider the value of live – rather than dead – animals before? Likely, partly because no-one has thought to do the calculations. But given that climate change and biodiversity collapse are new phenomena, there is now a rapidly growing movement among the largest companies in the world to invest in restoring nature not just to offset damage to the planet but to secure their own marketplace survival.

With an increasingly strict dependence on social responsibility, corporations are held to account by higher powers than government – be that their own insurers, investors or perhaps, now, nature itself. If this sense of responsibility is corrupted it creates conflict with and jeopardises their value to local communities who are partnered with those companies that are supposedly there to protect them. Our current political systems that value dead nature are far more corruptible. In the new economy people who jeopardise the future of elephants could be sued by the communities and companies responsible for their welfare.

Remember poor old Gus, the blue groper who met an untimely death in 'Grappling with Gropers'? No-one has ever put a value on a live blue groper even though they are responsible for the integrity of coastal reefs and protect homes and whole towns from falling into the sea. I can hazard a guess that the price of Gus alive and well would far exceed the $40 fishing licence that gives you the right to kill him.

In my first book, *Wildlife in the Balance*, I valued the ecosystem regulation services of a wedge-tailed eagle at $106,000. It was a simple and conservative calculation but enough to show that the absolute minimum worth of the bird is far more than the value society puts on it. Bear that in mind when considering a farmer in regional Victoria was fined just $2,500 for killing 406 of these majestic predators. If we consider that crime in relation to my calculation, he killed approximately $43 million worth of eagles.

Few people, especially natural scientists, frame these fundamental questions in a way that puts animals at the centre of the solution. Instead, policy often declares animals to be the problem, with their existence interpreted as a barrier to development. But as Ralph points out, in the economic sphere there is only a benefit to protecting wildlife.

In truth many people that live among abundant wildlife tend to be marginalised and poor and may not have the method or the means to consider the economic value of living creatures. Who knows what was going on with the farmer who killed those hundreds of eagles. The fisher who killed Gus may have simply thought it a way to get free food. It's easy to reprehend people for their actions but much harder to address the cause, which is that our marketplace doesn't value living animals over dead.

Even if most people want to behave better, how do we expect rural-living, poverty-stricken landowners to carry the financial burden of creating a whole new nature recovery economy?

Rural food gardens of villages in Africa and India can be annihilated by elephants. With farm soil health in rapid decline all over the world, animals and people are increasingly coming into conflict where any food and water remains, which tends to be where the elephants traditionally went. But we can mitigate that tension by better understanding the animals around us and coming up with cost-effective solutions that allow us to live side by side.

Enterprising owners of tea plantations in Darjeeling are signing up to the Certified Elephant Friendly™ tea program. Elephants mostly move through tea plantations at night following traditional routes. However, because tea isn't palatable to them, by preserving a permanent source of water surrounded by a couple of hectares of forest the elephants have a place to rest between their nocturnal journeys. Elephants can be tolerated moving through at night when workers are asleep and the refuges provide the elephants with a safe space to separate from people working the plantations during the day. It really is that simple.

Expanding efforts to build corridors between freshwater and enhance marginal land for biodiversity will inevitably improve the sustainability of a crop that's worth $17 billion. After all, elephants create the ecosystem and the value of one elephant is only a fraction of the profit of a business wanting to invest in conservation. The cost to a villager of an elephant destroying food, though, can be the difference between life or death. It's hardly a surprise the poorest people resort to harming elephants.

To shift the world's attention onto valuing animals we need recognition that their value is largest when they are alive. Then we need a method to channel money to people less fortunate, who we inevitably depend on to look after wildlife before we can transition back to a nature-based lifestyle. Critics of biodiversity offsets often overlook this need to invest where it counts. The very people worst affected and most allied to serve as custodians of wildlife are local and indigenous. Instead of offsets being used as an excuse for the wealthiest people to destroy nature, we can use the money to build more resilience for people and nature where it matters most.

If we can rebuild wildlife, reinforce beliefs in nature, make ecosystems more resilient and help people learn how dependent they are on animals, that will put us all in good stead for the next 100 years. This is where social and ecological scientists have a fantastic role to play.

LESSON 15: Wild animals are just as capable of taking lessons from us as we are from them.

Back in Bardiya National Park where about a third of Nepal's elephants live, only about one-fifth of the land is considered suitable for the current elephant population[80]. It will take time for the soil, water and vegetation processes in the land to heal under the terraforming influence of returning elephants, tigers, rhinos and countless other animals that together shaped the landscape and local culture in the past.

For now, elephants and people meet frequently and are having to relearn a way to understand each other to rebuild their relationship. Despite the clash of cultures, however, retaliation against elephants is thankfully quite rare. Sushila Chhetri, naturalist and guide in Bardiya National Park, says 'the only lasting solution is to accept the elephants and change the behaviour of the villagers' but admits the current compensation packages are insufficient. Electric fences don't work, and there is a tendency for crowds to form and rile the elephants with noise and rocks, which is the wrong approach borne of fear and the lost wisdom of cooperation.

Animals and people are intelligent enough to take lessons from each other. Over time, elephants and humans will learn to trust each other and give each other space. There have been no deaths since villages adopted the less confrontational approach with elephants. This passive relationship has also significantly reduced crop damage, meaning existing compensation funds are more likely to be sufficient[81]. Being simple truths these practices will spread between villages by word-of-mouth, better enabling the re-establishment of beneficial human-elephant relations elsewhere. This represents the dawn of a new animal-led, nature-based economy for the regions.

The people and tigers of Bardiya co-depend on the landscaping power of elephants and rhinos to maintain the water cycles that provide food, forest and shelter for all. When we finally saw our tiger in Bardiya she materialised at a riverbank, walked calmy into the water and bathed in full view. It was one of the most rewarding wildlife encounters of my life because she chose to reveal herself to us. She wasn't nervous and neither were we. We stood in plain view and she calmy rolled about in the water, licking her fur and paws.

There was no-one else in sight – no tourism vehicles, no noise, just the roar of another invisible tiger nearby, the barking calls of sambar deer and the ever-present 'pok pok pok' of coppersmith barbets. When she was finished bathing she walked away and dissolved into the forest once more. The value of that experience is undeniable.

Lessons from the Air

In his book *The Power of Trees*[82], Peter Wohlleben reminds us that trees, like animals, have their own culture and intelligence. It's a 'wow' moment when you realise trees can store their intelligence for thousands of years. Despite our mobile brains, we animals can't do that. Our culture is passed on through stories, song, dance, shared experiences and the ability for quite sophisticated communication between species using body language engendering mutual respect. Think of it as kind of a like a wi-fi signal that seems to require no physical connection to enable communication.

Compared to ancient trees our wisdom is ephemeral, and being an animal living inside one's own mind can be quite a lonely existence. Trees, on the other hand, use fungal networks like phone lines to permanently connect them with other trees all around and enable them to communicate.

Nevertheless, ancient trees in modern ecosystems are equally as dependent on wildlife as we are.

For a sedentary life form like a tree a 1,000 year-old brain is only good enough if the climate for their living remains stable. As Wohlleben says, 'In hot dry summers, trees have big problems. They cannot escape to the shade, and they cannot take a sip of water to cool themselves down.' Ancient trees need mobile, short-lived, fast-moving and adaptable wildlife rebuilding ecosystems around them that delivers predictable water and nutrition to a rich and diverse network of vegetation.

A surprise finding in recent research shows that the removal of one-third of the leaves of African desert trees increases sap flow, thereby further reducing the trees' thirst and putting more regular and predictable moisture back into the atmosphere[83]. These quenching mechanisms support people and all other animals too.

While we address the existential threat of climate change and move away from a fossil fuel dominated society, nature restoration is going to become even more important. Partly because natural climate, what I refer to as a 'fair climate for living', was the by-product of all the animal-led processes that create water, food and shelter. Animals condition the air, soil and water so that it becomes habitable. Removing carbon from the air doesn't necessarily replace those components of life support, particularly at a local scale where we need it most.

There is a risk that the focus on carbon as a pollutant will be a distraction from the importance of carbon in ecosystems. We can easily get tied up in technical solutions for drawing down fossil fuel carbon while we phase it out, forgetting that carbon is also the basis for all life on Earth. Carbon is food and we need less of it in the sky and more in the ground.

For example, we now know that elephants, whales and other megafauna have an enormous impact on the global climate. What about our local climate, though? If a few large animals can have planetary impact, how do many smaller animals affect the climate locally or otherwise? How much impact can they have? A lot, as it turns out.

As we have already discovered, relatively small declines in species abundance can have a greater effect than climate warming. Gavin Schmidt of NASA's Goddard Institute for Space Studies says we don't really know what's going on anymore[84]. Loss of wildlife and biodiversity is one key reason why climate scientists continue to struggle to predict the intensity of floods and bushfires. As they say, with every problem is an opportunity. In this case, extremes caused by the decline in powerful biodiversity feedback loops can be turned into opportunities to have an equally profound effect for good if we restore animal populations.

Masters of the Atmosphere

One of my earliest memories as a kid was waking up as the warm sun beamed through my curtains. Just as I was stirring from my sleep, house martins would be bringing the first bolus of protein-rich insect prey to their nest and would start feeding chattering chicks just beyond the window frame.

We'd often head to the supermarket in the nearby 'big smoke' of Bourton-on-the-Water in the Cotswolds. Overhead there'd be swarms of swifts darting across the sky. It was a magical time. Beneath the clouds was a world of wonder in flight and the air was full of natural sounds. At that point I never imagined I would need to ask, what's the point to swifts, swallows and martins? But here we are.

Thirty years later and our skies are falling silent, our minds filled instead with the noise of urban rumble. Though you might block out that pervasive low-frequency sound, it still exists and even increases our risk of dying from heart disease[85]. Our normal state of mind is one of peace that is only interrupted by the occasional sounds of nature, such as ever present bird song. The fate of birds critical to humanity is left for politicians to debate and people to decide whether to allow them to nest around our homes.

The wonderful activist and bird conservationist Hannah Bourne-Taylor failed in her first attempt to introduce nesting bricks for swifts into UK homes, partly because energy-efficient homes are increasingly airtight. You'd have thought that leaders inspired by such environmentally progressive ideals as energy efficiency might have a heart for wildlife too. Apparently not. The move was blocked by the outgoing Prime Minister Rishi Sunak in 2024. But swifts as a species are in danger of going nationally extinct. Whole species of birds can arrive from thousands of miles of migration to find their nest sites gone. That the UK Government has no desire to fix this is appalling – especially when the solution is as simple as letting these nest-perfect holes in the wall remain.

There are silver linings to all clouds, even if they can appear dark at times, and it's good to see the increasingly relentless fight from people to protect what they love. As we begin to understand how much more we've lost than just the animals themselves, we're also starting to realise how these losses affect our own survival in so many ways.

In Spain, two people have raised the money to install an artificial structure with enough space for 300 bird and bat nests, with swifts being among some of the species of birds they hope to attract. Diego Alves and Ángeles Mora's project is the kind of locally led creation that governments all over the world need to support[86]. Their project is in the town of Coria del Río just outside Seville in southern Spain. The town has been suffering from the effects of West Nile virus, a debilitating mosquito-borne disease that occurs throughout most of the world's tropics and subtropical zones. Climate change is making the region warmer and wetter, meaning more and more people are vulnerable to being bitten and infected.

The council could have used pesticides but the Mayor of Coria del Río, Modesto González, believes that spraying highly populated urban areas with toxic chemicals is unsatisfactory as it will further damage the environment (we learned this lesson in 'Farms with Teeth'). These sledgehammer tactics would end up killing more of the birds, bats and other insects that we need to keep mosquito populations under control. It is heartening to see that the town rejected this in favour of a much healthier nature-based approach.

As the number of birds increases, the risk of disease declines. Yes, you heard that right. Without birds to eat those pesky infection-ridden mosquitos we risk being bitten more frequently and seeing more people become ill. Disease mitigation is just one function that stems out of food security for birds. For example, migratory birds flood into areas to exploit surplus energy. In this case, that energy is in the form of carbon-based lifeforms called mosquitos. We need birds to appear in large numbers where mosquitos breed. There are other benefits as well, as our vulnerability to disease also increases when our food security declines. Our food security is reduced when we have fewer birds around.

In the UK, 3.5 trillion insects amounting to 3,200 tonnes of biomass migrate into the region annually. This aerial plankton is fed upon by birds and bats that concentrate, disperse and amplify nutrients from this food source. These fertile food-bowl floodplains that provide market garden vegetables to your local supermarket today were built on animal activity, the transfer and laying down and storing of soil nutrients for millennia.

Most of our cities are built on top of these places; Seville's famous market gardens are built on a fertile floodplain that was created long before human civilisation began. The insects (including mosquitos) and lesser wildlife such as bacteria represent a thinly distributed threat that humans don't have the resources to address before they become an impediment to our survival. Who saves us from this threat? Birds.

> **LESSON 16**: Migratory birds thread through wetland ecosystems, safeguarding water and protecting us from disease.

Furnished with this new way of thinking about the significance of wildlife, you soon realise that we share our fate with swallows, swifts, martins and more. Though my imagination and love for wildlife was sparked by seeing those birds as a child, it was only a few years ago that I began to realise how deeply that feeling is embedded in my survival instinct.

Birds have always been a representation of freedom and it is through their existence that we are free of so many of the things that might threaten our survival. Unless we're Superman we can't fly and that means our home ranges are small, so we depend on more mobile animals to create the habitable ecosystems that we know.

Before the invention of fossil-fuel-driven engines, human beings were mostly sedentary creatures. My grandfather barely left the valley in the Cotswolds where his ancestors had lived for generations. In stark contrast there are animals a fraction of our size that undertake the most incredible journeys each year. These are truly the world's best high-fliers. They are the most authentic travellers whose ancestors have overseen thousands of years of human civilisation develop.

The bar-tailed godwit undertakes an epic migration on a bewildering scale. Despite weighing only 300 g its home range encompasses a quarter of the planet. Each year it travels over 10,000 km and does this in a single flight at altitudes of 5,000 m across the Pacific. How? They navigate by internal compasses, read the stars and fly at an optimal height where the air is just thin enough. Sometimes they fly non-stop from Alaska to Australia or New Zealand but somehow know to deviate and avoid storms. Aviation engineers could learn a thing or two from these tiny adventurers.

Shorebird researchers started to put satellite trackers on bar-tailed godwits, which fly in a circuit around the Pacific between breeding and wintering grounds. Ongoing work by one of the paper's authors Adrian Riegen and the Miranda Shorebird Centre has allowed enthusiasts to watch these flights in almost real time. One individual labelled '4BBRW' holds the record for a single non-stop flight from Alaska to Tweed Heads on the north-east coast of Australia in September 2021. This distance of 13,050 km was flown in a little over 10 days.

To prepare for such flights, bar-tailed godwits alter themselves physically. Not only do they put on a huge amount of body fat that accounts for over half their weight but to accommodate this they shrink their gizzards, livers, kidneys and guts. Only the organs essential for long-distance flight are maintained. Then they rebuild their internal makeup upon arrival at their destination.

One can imagine the immense pressure this puts on every cell of their body as they live for weeks at the very edge of endurance. Each cell blasts out non-coding DNA fragments to reprogram a body with new proteins that respond to the threat of extreme weather, constantly changing altitude and being plunged into heat and cold. This enables their survival at the very edge of extreme conditions. And all for what? What do we learn from this incredible journey?

As mentioned a little earlier, these birds use a mental map for navigation that almost certainly includes star charts and beak compasses. Scientists have discovered nerves in migratory birds' beaks that contain iron[87]. They appear to be able to use these to monitor their precise position, even altitude, based on the direction and intensity of Earth's magnetic field. This is a superior way of navigating than a simple compass, it's more like an inbuilt GPS.

These birds don't so much 'read' these directions as 'know' where to go, as these are migrations that have evolved with the continents over many thousands of years. Surely something so unlikely cannot have evolved unless it was vitally important to the world's ecosystems?

The movement of abundant animals across Earth is based on behaviours developed over timescales longer than any of us can imagine. Like rivers, animals carve out ecosystems and create their own habitat along the way. In the case of the bar-tailed godwits, they have been nurturing wetland ecosystems since the last ice age 10,000 years ago. Since the beginning of evolution their ancestors have been learning to do similar things.

Migrating monarch butterflies do it, too. They remember the path around an ancient mountain range long since razed to the ground by glaciers, now lying invisible beneath Lake Superior. They don't fly directly south over the lake where those mountains used to be, instead for a short while they make a dogleg towards the east.

Kathryn Schulz writing for *The New Yorker* gives another sense of this ancient contribution to the planet: 'Ornithologists suspect bar-headed geese fly over Mt. Everest because they have been doing so since before it existed. When it began rising up from the land, some 60 million years ago, they simply moved upward with it'[88].

The godwits don't 'choose' to migrate, 'follow' a celestial map or 'decide' where to feed, instead they follow patterns of movement engrained into their culture that has evolved with the ecosystems they inhabit today. They are as much a part of our landscape as a rock that sits in a riverbed. But unlike rocks and plants, birds can move long distances so their floating brains travel much more of the Earth's surface than ours do. And metaphorically speaking they are our rock, the foundation stone for human civilisation. Their kind have held our coastlines together, shoring up entire civilisations, giving us food to eat and homes safe from the sea.

For years a proposed development at Toondah in Queensland, Australia was threatening to reclaim a coastal estuary and build 3,600 apartments and a 200-berth marina on top of a protected wetland. It was eventually defeated by a rambunctious rabble of birdwatchers, dog-walkers, cyclists, runners and all manner of local people who use the coastline. The bar-tailed godwit was one of the signature long-distance migrants that collectively defined the 'character' of wetlands as described under the Ramsar Convention. This international policy aims to protect a global network of wetlands because they are essential for all of humanity. The fact these birds still occurred in relatively small numbers on the site signified that the ecosystem's overall integrity was still somewhat intact and, therefore, salvageable.

BirdLife Australia's lead campaigner and chair of the Toondah Alliance Judith Hoyle says, 'At first, people didn't understand [this species'] importance. The developers had a panel of independent scientific experts who said there would be no impact.' This misinformation reassured members of the public.

Eventually it took 11 years for the campaign to reach a head. Judith says,

> It was like a full-time job. The Toondah Alliance had community art events, Mother's Day marches, art exhibitions, postcards and quilt making, people emailing MPs and so on. We changed a trickle to a river and then a tidal wave. The momentum resulted in people staying the course and when the time came to do their formal submission, by God they did!

In the end this campaign of information won out, and there were 26,000 submissions against the development, amounting to 85 per cent opposing it.

Had it not been for Toondah Alliance's determined campaign, this could have easily been an unhappy ending for our feathered friends and for people in the local area. The first problem was that the developer was permitted to argue that a massive decline in waterbirds would represent an opportunity for the local community as their consultants let them believe that the wetland had more capacity to be developed because there were now fewer birds. This would only be true, of course, if it wasn't the birds creating the habitat in the first place.

'People now understand that poor science was responsible,' says Judith. 'Unless you have a well-organised system to rebut this [poor science], it ends up costing a lot more.'

This is another vital lesson that we must learn: don't have faith that everything scientists tell us is fact. Ironically, had the developers not shone a light on the real value of the wetland the threat to it may have remained invisible for longer. This serial incompetence we have where we make poor decisions and learn from them later is arguably the only reason we know anything about the environment. Our society is rather immature, naive and lacking in both knowledge and beliefs (which is why we need this book) though, ironically, this could end up being a blessing.

Any belief system based on fundamentally poor knowledge can turn out to be very expensive. In this case, a multi-million-dollar development was given conditional approval. A nod from the government in charge and some ecologically creative accounting of numbers of birds meant no-one seriously alerted them to the pitfalls in their argument. Look deeper and one might find a case that the advice was somewhat negligent, but it is doubtful anyone will be held to account.

In truth, many studies of this kind are administered at the whim of a client. Very good consultants can be given the narrowest of briefs, making them unable to appropriately advise on the best course of action. This is an argument for stronger regulations of the environmental consulting profession. In the way that doctors heed the Hippocratic Oath and lawyers uphold justice, ecological consultants should uphold the rights of nature to exist because of its stark economic value. However, the lack of a simple and coherent belief system and laws that ensure consultants act in nature's best interest means the strongest advocates for ecosystems tend to become the least likely to be employed as advisors. This leads to inherent economic downfall as decisions get made that ignore the value people have for nature and the value of nature to the economy.

Little more than a dozen out of more than 1,700 federal environmental assessments for developments in important habitats have ever been turned down, suggesting the odds are stacked in favour of the developer and against wildlife. One reason such sensibility has not triumphed yet is that consultant scientists are allowed to set their own agenda rather than providing answers to the community. It's strange that public submissions for advice and feedback are received at the very end of the process. A wise consultant would assemble the concerns throughout, address these and the final submissions would only represent the residual concerns. However, that would mean compromising on natural capital impacts and since the question 'how much is the ecosystem worth?' has never been asked, this process creates inevitable conflict.

Wherever and however community-designed protection happens, the fact is it bolsters communities. In her book *The Reindeer Chronicles*[89], Judith Schwartz relays a story by Jeff Goebel of the Community Consensus Institute. Jeff helps communities remove obstacles so they can reach their own goals. In Mali, Jeff explained, he brought several warring ethnic groups together who were facing critical food insecurity. Reframing the discussion around what *could* be possible, he invited a new style of conversation. The local people rose above a stalemate and exceeded their own expectations.

'It was their set of beliefs that convinced them it was impossible,' he says. 'Once we acknowledged it could happen, it took the pressure off. This freed up their minds so they could consider what they *would* do.'

Very little work has ever been done on the value of shorebirds to the health and integrity of coastlines. Given their vast numbers and omnipresence, shorebirds are clearly a major player. Search the scientific literature, however, and you'll find an abundance of studies on how shorebirds are dependent on estuaries but nothing on how estuaries are dependent on shorebirds.

As with Rosling's experiments with people and chimps in 'Lessons from our Past and Present', it will take years for this significance to enter the general psyche. Evidence of this was repeated as recently as September 2024 when the Scottish Association for Marine Science (on behalf of WWF, The Wildlife Trusts and The Royal Society for the Protection of Birds) announced a landmark study[90]. They estimate that the carbon sequestration potential of UK wetlands including mudflats is almost three times that of its forests. Most the carbon is stored in the top 10 cm. Guess where all the birds, animals and microbes mainly feed?

Chief correspondent Alex Thomson for the UK's Channel 4 News broke the story saying, 'Such numbers [of birds] are supported of course by the richness of that mud again – its worms and invertebrates one giant bird feeder for them during the day'[91].

There is a principle in science called Simpson's Paradox where the trend between two variables can be reversed when a third variable is added, such as when you make assumptions about ecology at one point then introduce a bigger scale. For example, we might be able to afford and get value from using pesticides at the small scale, but it collapses at the large scale (as we discussed in 'Farms with Teeth'). Likewise, when we remove animals we see vegetation quantity bounce back but given more time this collapses whole ecosystems (as we discovered in 'Ghost Builders of Borneo').

Likewise, our view of shorebirds is limited. To understand complex systems we unsurprisingly break them into small units. After all, humans have naturally narrow vision. We see birds feeding *on* an estuary and therefore consider their impact *on* that estuary; science compartmentalises ecology at a moment in time and place.

The fact is migratory birds are a significant reason as to why that carbon is stored. Not only that (as we will discover in 'The Fear Effect and Food') but their actions enable microbes to massively increase that rate of storage, which is fundamental to our own physical and mental wellbeing. Over millennia, it's not the animals that are feeding *on* the habitat, it's the specific individuals most likely to contribute to building a self-sustaining ecosystem that enable this self-fulfilling natural infrastructure to evolve. This continues providing more and more animals with life support. It's the opposite of what we think.

Why is this important? Because as Paterson says, mudflats are 'a huge asset for decision-makers. Now we need them to act on its findings'. How can you protect ecosystems until you realise the critical need to put wildlife first?

The global population of bar-tailed godwits today is estimated to be just over one million birds, even though they may have declined by at least half since the 1980s. Their epic flights see them arrive in vast mercurial flocks where they probe deeply into mud layers using a lance-like beak. They land in droves, their movements mirroring the ebb and flow of each tide as they use instinct and culture to map and navigate to areas of plenty. The species has performed this ritual trillions of times over millions of hectares, over tens of thousands of years and longer.

Estuaries are covered with biofilm, which is a nutritious, energy-rich layer of diatoms (microorganisms full of fatty acids that excrete a biological glue). These sticky films mix with the surface up to a few centimetres deep and hold it together[92]. The smaller and more numerous birds – the stints and sandpipers – lap at the biofilm with feathery tongues. On twice-daily tides, before every individual of these many millions of birds lift off, they each deposit some of their weight as guano. As the first waves begin to pass across the estuary, hundreds of kilos of this nutrient are spread gossamer-thin across the mud as worms, hermit crabs and shellfish lap it up and secrete it into the sediment layers.

The birds' activity ensures these biofilms are constantly topped up. They could be among the most productive ecosystems on Earth – we don't know. The diatoms' gluey waste holds fast in the tides, giving seagrass – upon which fish make a living – the chance to take hold. As these roots grow deeper, saltmarsh or mangroves might appear. Where the birds are most abundant you can be sure the opportunities and risk are both extreme, as these places tend to be outstandingly dynamic. Roebuck Bay in the north of Western Australia, The Wash in the UK, the Yellow Sea in China and Humboldt Bay in California are just a few that spring to mind.

It's not just these places, though, it's also everywhere in between. A common misconception among consultants looking at shorebirds is that simple numbers of birds have some kind of meaning. Consider this: you could count a maximum of 15 birds in one day of a three-month-long season and theoretically have 1,350 birds passing through. That 15 you count are only the 15 that fly through on that day. If we take that example then we never know exactly how important a site is, just that it is important because it has birds. We might decide to use the maximum count instead or compare it to other places. Who is to say, however, that fewer birds in one populated place is of less value to those local people than many more birds somewhere remote? Those birds might also be the most significant breeders and the smaller urban site is the only reason the population is stable.

Just because we don't know that birds are essential for our economy doesn't make it less likely that they are. Again, what we don't know about nature more than outweighs what we do. The question is, what do we do about that? This is something we will come to in the closing chapters of this book.

Either end of their 13,500 km journey, bar-tailed godwits fan out across the estuaries of Australia and New Zealand and some of them visit multiple places. As they deplete their prey in one place, they move onto the next. These periods of plenty and recovery are the beating heart of ecosystems across the world.

Global tracking studies, while wonderfully interesting, represent only the most extreme dimension of their behaviour, so our knowledge of these amazing creatures is only partially complete. Like the effect of tides on coastal erosion, it's not the '100-year storm' but the daily grind and the nondescript – often invisible – actions of birds and animals hiding in plain sight and sharing knowledge of where to feed throughout the season that amounts to the greatest impact on our shores.

Suddenly it seems more reasonable that we should accommodate these animals among our own living spaces and not build on their homes. Leaving a hole in the wall for swifts and martins is a small price to pay to maintain the rich ecosystems that support our lives. Their survival is so strongly connected to our own, it only makes sense that we should give them a right to exist.

Those of us who don't live on an estuary might consider giving our friendly neighbourhood swallows, swifts and martins a break and offering them a nesting site. Still not convinced? You should be. Our increasing urban lifestyle comes at a cost and their certainty to arrive and breed every year in spring is the nudge we need from nature to remind ourselves about the lessons they teach us for survival.

We might only glimpse these high-flying friends for a second each day but with them we regain our own freedom to live more easily and healthily. If, like me, you can't imagine a world without birds, then it's time to make your voice heard and join those who protect them with all your might.

Billionaire Big Dreams Versus Reality

On stage at the 2023 Sustainable Development Goals Summit in New York, Bill Gates said it was 'complete nonsense' to think planting trees could solve climate change.

Naturally, Gates had a lot of people worried, especially as he speaks from the position of being a billionaire technocrat from the hardware and software space who holds sway among many nature-deficient and urbanised businesspeople. Salesforce CEO Marc Benioff, on the other hand, has gone to the other extreme and thinks we should plant a trillion trees. In his view, this is how we solve the climate crisis. Both have merit but neither will work at scale because even tree planting is an engineering solution, it's a human intervention in the delicate balance of our ecosystems that is not ecologically aware.

Why are our planet's movers and shakers so obsessed with trees? Because trees are massive and it's obvious that they are made of a lot of carbon. And carbon is the earth's thermostat regulating our temperatures. It's also in the food we eat and, as I'm sure those billionaires all know, it forms the energy that fuels our whole economy. But as we've seen in the 'The Surprising Elephant Economy', trees aren't the only things that are important for carbon. Size isn't always important, either.

We spent several hours in dunes on the Great Ocean Road in Victoria, Australia. It was six degrees overnight, but by 10am the sun was lightly baking the ground. A soothing sea breeze was lifting moisture off the waves cresting over the beach nearby. Leaning forward to inspect a bush no higher than our ankles we immediately felt a change in humidity. The sand became slightly browner as it accumulated organic particles and retained heat and moisture. Not every habitat is identical, though, and it takes a trained mind to sense the right conditions and locations where tiny spiders thrive.

We were looking for peacock spiders, which are a type of jumping spider. These bonny arachnids aren't much bigger than a grain of rice and most species were only recently discovered. The name comes from the performance put on by some males, an elaborate dance where they fan their 'tail' to reveal stunning coloured patterns. But they're not just a pretty face – they are also ferocious hunters. We find one sitting on top of a patch of vegetation no bigger than coffee table. Despite the small size of that area it's incredibly diverse, boasting perhaps 20 to 30 types of plant.

Recent studies have shown that hunting spiders – think wolf spiders or jumping spiders – are essential to carbon cycles. This is something that Professor Oswald Schmitz at the Yale School of the Environment has been trying to get the scientific community to pay attention to for a decade or more. In his 2017 study in the journal *Ecology*, his team showed that the most carbon retention occurred when there were active hunting predators rather than sit-and-wait predators present. By that, they mean the ones that run around and seek out their prey rather than sit and wait at their web.

So how do spiders increase carbon retention? Trees need to remain in the ground permanently and for a long time to sequester significant carbon. A tree that falls, rots or burns releases this back into the atmosphere. Animals offer a far more effective way to store carbon regeneratively because without them vegetation can't diversify. The amount of carbon isn't maximised or used in processes beyond climate. Rather than the carbon just sitting there, wildlife creates its own market by drawing in other wildlife that amplifies the carbon sequestration processes and creates a measurable dividend that goes way beyond mitigating climate risk.

The presence of a spider predator forces the creation of ecosystem structures at the microscale, maximising carbon retention and all its benefits. There is a generally accepted principle in ecology that systems are regulated from the top down, but not as we might expect. The regulation is top down *throughout*. As we discovered in 'Farms with Teeth', anything that eats something else is a factor in the stability of the systems beneath it. And as we will discover in 'The Fear Effect', life support collapses without predators and affects everything from our own health and nutrition to climate. The biggest animals at the top might hold the greatest significance in humanity's perception, but nothing can exist alone. Schmitz's work establishes that even the tiniest creatures can collectively transform our world in ways we're only just beginning to imagine. Spiders consume from 400 to 800 million tonnes of prey every year from forests and grasslands[93] and because of the sheer abundance of them, this adds up to a habitable planet.

The ecosystem we were seeing on the sand around our feet that day in the dunes wasn't made by plants. It was moulded, broken, lived in, eaten and rotted by animals, some even comparable in size to us. An audible 'thump-thump' reveals one such creature – a swamp wallaby lurking nearby, grazing on spring's new growth.

But the overall shape of things is best when there is a range of wildlife in balance, all different sizes and with a diversity of roles to play. As we should know by now (I'd hope so, being this far through the book) people are part of the same system and everything we have today was built on the impact of all animals working together.

When Europeans first arrived Australia's soils were a metre deep, soft, crumbly and spongy. This life-support legacy created by people and other animals is mostly gone. Today the forests that have fallen silent teeter on the brink. They are tinder dry, repel the ocean's daily outbursts of moisture and burn with unprecedented ferocity.

A fair climate means water in the right place at the right time. Farming and fisheries depend on knowing when certain seasons come and go. While there is the problem of global atmospheric carbon accumulation from fossil fuels, we also need nature and wildlife to create a fair, balanced climate for living locally. And that means allowing wildlife to put carbon where it's useful for life support – brought into the bodies of animals, mulched into soil and built into diverse and species-rich grassland and forests – rather than drifting into the atmosphere or acidifying our oceans.

LESSON 17: The answer is not simply more trees, but more wildlife-rich habitat.

By reintroducing eastern barred bandicoots, the Odonata Foundation's Tiverton project has proved the restoration of soil and water cycles on farmland. Planting trees couldn't have achieved this outcome. Like the sea urchin-dominated underwater plains in Komodo, these are pioneering land habitats that were the lifeblood of humans for thousands of years.

Peter Guiden and colleagues at Northern Illinois University found that 'the effects of ... animal communities were six times stronger on average than the effects of plant biodiversity'[94]. There used to be 50 million American bison grazing the plains of North America. European bison are now being reintroduced all over Europe to redress soil decline. Grasslands are highly dynamic ecosystems that require a particularly acute level of animal interaction to hold together, but much of that is done at the local scale.

As an example, earthworm contribution to ecosystems makes 140 million tonnes per year of global food crops; that's about 6.5 per cent of global grain production[95]. Earthworms can't exist without soil and soil can't exist without animals. Where else does that rich, mulchy, organic material come from? It can't be just from plants because we've been farming with raw ingredients in the form of farm fertilisers for decades and all that's done is strip soil of its nutrients. About one-third of the world is naturally grassland and when it's looked after it can have 7,500 times higher diversity of plant species per square metre than an equivalent area of South American rainforest[96]. This astounding level of complexity is removed when we farm trees for carbon. We need animals to make trees into forests, and we need forests to be one part of a more balanced landscape held together by animals applying top-down pressure all the way from wolves to wolf spiders.

The United Nations recognises the dual problem of biodiversity and climate as 'two sides of the same coin'. I'd go further and say that while atmospheric climate change is a fossil fuel problem, a fair climate for humanity and other animals has become a wildlife-restoration problem. The cumulative, small-scale effect animals have on building a habitable planet all adds up to a fair climate. Though many people in Western nations remain reluctant to put animals and people before plants.

Gates, Benioff and many others live and work in cities. With drivers for change coming mostly from European and US-led policy decisions, a chasm has formed between the privileged opinions of a few nature-deficient people and the rich wisdom of first and second nation local people who retain a connection to their country. Gates is right to say that planting trees won't solve climate change. But Benioff is also wrong to say it will.

Neither have offered up the real solution, which is to stop burning fossil fuels while also restoring wildlife populations.

That is the most important lesson here. It is not enough to just stop polluting our planet; we need to protect what we have and rebuild on the margins, allowing animals the autonomy to roam wild and recolonise. And by that I don't just mean endangered species, I mean everything from elephants and whales to jumping spiders and aardvarks. The quality of our air depends on this and that means humanity does, too.

It's not just about our atmosphere, either. Nature's services extend to flood and drought reduction, clean water and pollination, all of which are thought to be worth over $125 trillion globally. That's more than one and a half times global Gross Domestic Product. This is why GDP is a terrible measure of a civilisation's efficiency, because it is 'fishing from the bottom of the barrel' and it's not sustainable as it doesn't value either the barrel or the live fish. The European Union's Nature Restoration Laws that came into force mid-June 2024 expect eight to 38 times return on investment for every dollar spent[64].

When wildlife is involved, carbon is permanently locked away underground and stays there despite what might happen at the surface. You could conceivably build on top of soil without destroying your investment. The same can't be said for trees that would have to be cut down. As beautiful as trees are, planting a trillion of them won't bring back the balance of our planet's climate. Instead, getting the balance between wildlife, forests and other habitats right is essential in the fight against climate change.

Clean Air and the Cost of Hayfever

What's a fair climate if it isn't the chance to step outside, take a deep breath and … 'Achoo!' Excuse me … sniffles … rubs eyes … oh dear. It's back.

Every year unseasonal rainfall and warmer-than-average weather seems to lead to a massive growing season for grasses and that means pollen galore. For hayfever sufferers like me it can be a miserable time, but it's only since moving closer to the coast I've realised how much discomfort I've been putting up with for so long.

Here a sea breeze pushes the particles inland and I feel a lot better. But pollen isn't the only airborne danger that we need to be aware of. I used to live on the brow of a mountain, and when I drove into the city, I could see the smog layer rising into the cold air above. Smog is where a temperature inversion occurs that means warmer air gets trapped at ground level. When it rises against the cold, it layers concentrated pollutants over vast suburbs. This, too, is a hotbed for allergy sufferers.

Many of you might remember a thunderstorm that swept in from the north to Melbourne in November 2016. Vast quantities of grass pollen from the northern deserts of Victoria were swept up in this tempest and dusted into the city. Why was this storm so significant? Because this one almost collapsed Melbourne's health system due to a new phenomenon – thunderstorm asthma. It was deadly, killing 10 people as thousands more were admitted to hospital with respiratory problems.

But these occasional events aren't our biggest problem when it comes to the quality of our air. Hayfever costs Australians nearly $30 billion a year in healthcare, days off and reduced quality of life[97]. It's another example of a major public health problem in developing countries that has crept in slowly. Most people endure low-level 'chronic' symptoms, but it's only when we look at these wider social consequences of pressure on healthcare providers and economic burden that we realise just how out of control it's getting.

Is this to do with climate change? It's certainly a factor, but there is another cause that has rarely been considered as (once again) science hasn't asked the question. This could be a result of wildlife collapse. We've already learned that small declines in the abundance of wild animals amounts to a greater risk for society than most known threats like nutrient pollution and climate warming. Plummeting global insect and herbivore numbers could be contributing to the huge excess of pollen.

Most of us are aware of the role of insects in pollination. What we hardly talk about, though, is the role of pollen in ecosystems and particularly its consumption by animals. These creatures are called 'palynivores'. What you might not know is that most animals eat pollen and this has been going on since flowering plants evolved about 300 million years ago. Research has found ancient evidence of pollen consumption by flying insects, springtails and mites[98].

When we think about things that consume pollen your first thought is probably bees, but they are just one of a wide range of creatures who love a dinner of this nutrient-rich food. Ants eat pollen, too, possibly far more than we imagine. They deliver spent pollen to the springtails, fungi and other soil fauna and flora by burying waste underground. Consider, also, that ants are among the most widespread, abundant and diverse groups of insects on Earth.

If we move up the scale in size then bees consume pollen on an epic scale, with the average honeybee colony putting away a whopping 40 kilos a year. Spiders also eat pollen and do more with it than just supplement their diet. For our arachnid friends, pollen increases spiderling survival by making up most of their food. Grasshoppers and beetles all eat it, and for specialists like pollen beetles it's their entire diet.

This is the point where things might get a little unexpected because larger herbivores such as bison in the US and Europe or kangaroos in Australia (which declined to a fraction of their original numbers) would once have helped keep pollen under control, too. They not only do this by munching on vegetation but also by introducing soil nutrients through defecating. The physical impact of hooves creates structure at the ground level that to tiny jumping spiders and grasshoppers is like immediately growing mountains and fertile floodplains right across the landscape. This diversifies plant and animal life in the soil and grassland, vastly increasing the number of pollen-eating insects.

Pollen is consumed on a massive scale because it is really protein rich. It's not uncommon for animals to switch their diet, especially for their offspring, to a protein-rich alternative and pollen does that job very well.

It mostly erupts in springtime at the perfect moment to be of greatest benefit to the whole food chain. After all, it is one-half of what's needed to create new life.

Pollen is plant sperm and, just like in animals, it is produced in vast excess. Plants produce millions of tonnes of pollen every year that can be swept around the Earth. It was a significant component of a storm in Northern Europe in 1991 that deposited 50,000 tonnes of it on the land. Each square centimetre contained up to 1,170 pollen grains, which was enough to turn the snow yellow.

German botanist Hans Ferdinand Linskens notes, 'This impact that pollen has on plants, animals, and man is the direct consequence of its mass production'[99]. Single alder trees are capable of producing seven billion pollen grains each season, amounting to tonnes of pollen from a single street in Buchs, Switzerland[100]. One hectare of pumpkins can produce 160 kg of the stuff. The lifetime pollen production of a single beech tree has been estimated at 20.5 billion pollen grains, while another study estimated nearly 13 billion grains per square metre of mixed vegetation. In 1969 Linskens describes how even 50 km off a European coastline 8.8 to 16.2 pollen grains were estimated to be present per square millimetre of air every 24 hours.

Pollen isn't just an important food source for air and land dwellers but also for the wildlife in our oceans. The pollen from pine trees dominates deep ocean trench sediment off New Zealand, which is extraordinary given that pine plantations have only been around since European settlement. It's even more surprising to think that this pollen could be an important source of food for animals like shrimps and limpets that graze sediments in the deepest ocean.

> **LESSON 18: Protecting our grassland insects and grazers lets us avoid chronic hayfever.**

When it comes to natural particulates, pollen is one of the main constituents of air that enters your lungs when you breathe. Hayfever doesn't just affect humans – dogs, cats, horses and other animals can also get it, too. But where does it fit into the wider scheme of things?

When you have something microscopic produced in excess, animals evolve to capitalise on this to keep the ecosystem in balance. As with mouse plagues, outbreaks of starfish and other overabundance of animals, it only takes a small change in an outside variable to create an imbalance that results in a Herculean eruption.

Remove ground herbivores and insects and we create the perfect gap in the market for the most highly virile plants to breed on previously wildlife-rich grassland. More plants mean more pollen. Take, for instance, those bright yellow oilseed rape fields that are notorious pollen producers. We spray nasty insecticide all over them to kill pollen beetles and without them it doesn't take long before there is nowhere for the excess pollen to go but into your lungs.

If we equate the integrity of our soil to the food we eat and our general health, why not equate the climate to the air we breathe? It should come as no surprise that our health is widely linked to the quality of our air. Once again, the solution to our woes lies in wildlife. Once we recognise the enormous value of wildlife for our own health, we will start to feel better.

Pollen is also a source of essential protein food for most insects. The large quantity of carbon this contains is transferred into the soil by animals to become the entire basis for global nutrition. We'll explore how this connects to our everyday lives in 'The Fear Effect'. Decline in grassland wildlife has led to an epidemic of surplus pollen that reduces soil carbon, destabilises climate, kills soil productivity and reduces water retention. Nature reminds us about this every time we sneeze!

Seabird Secret Sauce

Have you ever stood taking in the sea air on a calm summer day at a wonderful spot like Flamborough Head on the east coast of the UK, looking out from the cliffs and marvelling at the beauty of nature? Unfortunately, at this spot there'd be little that's very peaceful at all, as these cliffs are home to a 200,000-strong seabird colony of mostly gannets, guillemots, razorbills and kittiwakes. And in a place this populated the cacophony of bird calls from below is quite the experience.

If you look out towards the horizon from any coastline there is a sharp distinction between the land, the sea and the sky. In this moment, you'd be forgiven for thinking these are all separate pieces – individual elements of our coastal landscapes – and for humans they may be. After all we don't fly and are quite ill-equipped to swim, so any real connection we boast with the air and water is mostly experienced while contained inside metal machinery. Despite our ubiquity and mobility as a species, we barely touch the world beyond our feet. For seabirds there is no such mystery because these birds cross those boundaries. They are masters of land, air and sea, which makes them among the most important components of our planet's weather systems.

But back to our scene of relative tranquillity for a moment. A sea breeze ensues as the land starts to warm, then small clouds appear and the wind wafts upwards over the cliffs. Gannets use it to lift off and will soon penetrate that boundary between air and sea with their spear-like shape to feast on the fish life below the water's surface.

Then our moment of tranquillity is interrupted, but not by bird cries this time. Instead, a pungent odour is sent up our nostrils. The white streaks that adorn the rock faces below precipitous seabird-nesting platforms were dampened by a light rain last night and the guano – bird poo – has been volatilised into ammonia and sent skyward.

This smell is the by-product of an important ecosystem process and often it is so acute that many people have a visceral reaction to it. But this distracting smell denotes a substance that may be among one of the most important elements of ecosystem processes to come from seabird colonies. The climate for our farming and fisheries depends on it.

In the 19th century, guano was mined as fertiliser and may well have kickstarted the Industrial Revolution[101]. But it isn't just the land that benefits from this wonderful waste. Given that our nation's seabirds are also the biggest point sources of the gas that is ammonia on Earth then they also fertilise the atmosphere, which means – believe it or not – they make the weather[102].

During summer the ammonia from guano and ocean algae engage in a chemical reaction that creates sulphuric acid. These acidic droplets make clouds and regulate regional rainfall patterns. This reaction also significantly reduces atmospheric temperatures[103]. If you've heard of cloud seeding you'll know that chemicals can be flown into the atmosphere and released by aircraft to make rain. Either this is used to promote rainfall in each place or suppress it further afield. Remarkably, this process only takes tiny quantities of chemicals. In Europe, just nine grams of silver iodide an hour was enough to promote rainfall and halve the intensity of subsequent hailstorms.

The Chinese used artificial cloud seeding to ensure rain fell before the Beijing Olympics in 2008 so that it wouldn't rain during the games. When the country has previously used this fascinating technology, it has managed to influence the creation of 16 million metric tonnes of snowfall.

Ammonia has a different chemistry but is produced in huge quantities and emitted slowly and more precisely. The patterns of atmospheric ammonia production are in keeping with our needs because it's produced where and when we expect it to. Seabird colonies can emit as much as 90 kg per hour after light rainfall and total emissions from UK seabird colonies are about 2,700 tonnes a year[104]. Ammonia, and by that extension the existence of seabirds and their guano, is a vital part of our existence. We've even built a farming system around its effect.

LESSON 19: Seabirds link land, sea and sky making conservation, farming and fisheries co-dependent.

Farmers need predictable rainfall so they know where and when to grow crops, but for those ecological processes to align with cultural knowledge and become compatible takes thousands of years of cultural evolution. Very few communities around the world have learned how to use seabirds to their advantage.

On the island of Nauru in the South Pacific, local people have a tradition of taming frigatebirds to return to breed in front of their village. By the late 1960s, Nauru was the second-wealthiest country in the world from mining the guano deposits from its seabirds. When those resources – and perhaps the knowledge of how to maintain them – were exhausted, the island had descended into poverty by the early part of this century. These days this small nation has turned its gaze to the sea once more to make a living. We know that when seabird numbers return, this markedly increases fish density. Nauru villagers who rely on fish for food are enhancing their chances of survival again by making sure they look after the seabirds.

On Iceland's Westman Islands, people in the Vestmannaeyjar community honour an annual tradition where they go out at night in the months of August and September to rescue baby puffins or 'pufflings' distracted by streetlights. Thinking it's the moon guiding them to the sea, these baby birds end up in perilous places around town a long way from their destination. The next morning, kids launch them off the cliffs and back to where they belong.

Puffins have long been an important source of protein for people living on remote islands in the North Atlantic. According to interviews with locals by the *Smithsonian Magazine*[105] the tradition of rescuing pufflings started at least 90 years ago, probably with the advent of electric light. Likely at some point people became worried that an important source of sustenance was becoming endangered.

Today puffin hunting still occurs but is strictly regulated. The broader community's cultural connection, however, is maintained through the seasonal rituals of rescue and release of these fluffy bundles. The survival of puffins and people here is inextricably combined, leading to a stable ecosystem based around animals (as we also saw in 'The Budj Bim'). This connection to nature that Icelandic islanders hold onto is every bit the same as the connections any of us form with animals – wild or domestic – that we explored in 'Sympathetic Stingrays and Adventitious Apes'.

We should all be making every effort to protect and connect with seabirds, especially given that much of the human population lives near the coast. Seabirds are a huge workforce flying to and from ocean hotspots, centralising land, sea and air chemistry where it's most useful. Our choices of where to live and grow food were based around these patterns.

Like the birds we visited in 'Masters of the Atmosphere', we have aligned our movements and expectations with the landscape on which our civilisation formed. If this suddenly changes – either because overfishing means seabirds can no longer create predictable rainfall or we decide to tinker with the weather using aircraft – we further break our connection to nature.

While our technology can theoretically replicate this climate chemistry (as the efforts of China and Europe have shown) we cannot do it precisely. Instead, our attempts will chaotically, radically and temporarily alter weather patterns by breaking these natural processes resulting in a less fair, more unstable climate. Those age-old animal and human systems of cultural knowledge will disintegrate further, doing more damage than we solve. We must face the fact we cannot create a sustainable outcome without seabirds.

Transport of nutrients by seabirds is a huge component in global ocean and atmospheric processes, and where once seabirds and migrant fish such as eels and salmon would have moved about 150 million kg of phosphorous, that figure is now only a few percent of historic values.

Our survival depends on seabirds because humanity's basic life support – our food systems and water – are at stake if we allow their colonies to collapse. Keeping them in healthy numbers is not an impossible task. The number of puffins on the island of Lundy in the UK has increased from a handful in the 1990s to over 1,500 today. We can make this change and we should make it now.

The once proud nation of Nauru was brought to its knees by poverty long ago but it is perhaps the greatest irony that they might be among the most likely humans to survive. Why? Because some of their population started rebuilding their nature-based existence 50 or more years ago – around the same time as we began to destroy ours. There is another profound difference between our Western civilisations and their indigenous ones in that they retain their customs and connection to nature. I find strange the reports that their customs and relationship to frigatebirds have never made a connection with a culture of survival. Being unaware of this power is maybe their greatest asset. Armed with that knowledge it could be manipulated or easily capitalised. Instead, they are served well simply by living the way they always have, connected to nature.

The link between seabird guano and rainfall is one of many reasons why there has been a strong connection between world farming and wildlife since the beginning of industrialisation. Now's the time to instil a new belief system where farmers covet the benefits animals like seabirds still bring in the same way the people of Nauru relate their own personal status to the frigatebirds around them. Once farmers (and fishers, too) understand the true value of seabirds, conservation will turn a corner. We'll be able to begin saving animals from extinction, preserving seabird and human societies to ensure a long and happy future for both.

Lessons from Giants

If I have seen further, it is by standing on the shoulders of giants.
ISAAC NEWTON

Three thousand people fell silent as a gently smiling lady walked onto the stage of the Palais Theatre in St Kilda, Melbourne. With all the grace befitting a giant of conservation, Dr Jane Goodall took a seat in front of her adoring audience. In the year of her 90th birthday, Jane had come to address everyone in the packed auditorium on her 'Reasons for Hope' tour.

There are plenty of younger speakers who can fill stadiums, some of whom portray the most appalling characteristics of human behaviour. How many of them will still be able to hold the attention of an audience at 90 years old? If nothing else, Jane is proof that we still look up to the elders in our community. People like her have always been there to nourish and protect us with their wisdom and their voices endure even through times when they are in danger of being drowned out by an increasingly angry and divided society.

The wisdom of the natural world recognises that human and animal behaviour will be aggressive from time to time because, as we have seen, we cannot survive without conflict. But those who are truly wise and who offer hope know and accept this while still being supportive, encouraging and rich in the best of human endeavour. Jane Goodall is one such person.

Jane likely owes some of her wisdom to her mother, Vanne, who despite having no scientific background herself supported Jane's passion for nature well before her life was transformed by meeting Louis Leakey. A famous palaeontologist, Leakey was responsible for discovering the connection between apes, hominins and other early human ancestors. He was convinced that proving a similarity in the behaviour of chimpanzees and humans would add support for his theories of human evolution. With no formal scientific background and barely two coins to rub together, Jane and her mother lived in a tent in the jungle for six months to study chimpanzees. It was towards the end of this research that the young Jane made a discovery that was to become the trigger for her life's work and wisdom.

The first chimpanzee to trust her presence was a male she dubbed Greybeard. He would strip the leaves off twigs and use them to extract termites to eat. 'It was the first time any human had discovered tool use in an animal,' says Jane. 'Until then, tool use was considered a uniquely human trait. Indeed, humans were classified as "man the toolmaker".'

On presenting these findings to Leakey he joked, 'Well, I guess we are either going to have to redefine "tools" and redefine "man" or accept chimpanzees as humans!' This moment redefined humanity. Soon after this discovery, Leakey invited Jane to Cambridge University and negotiated postdoctoral funding for her. 'There's no time to do a degree,' he said. 'We need to get this information out there.'

Louis Leakey must have been an extraordinary man. At a time when women were given little assistance, he supported three burgeoning female scientists in pursuing their own fervent intention to study primates: Jane

Goodall, Dian Fossey and Biruté Galdikas. Together they became known as the 'Trimates'. Leakey knew that women, not men, would have more success habituating among primates and that their patience, resilience and open-mindedness would allow them to succeed where others had and would fail.

There may never again be three more important people in the world of conservation. One can't help feeling a bit of a fraud for telling stories about animals, wisdom and nature when following in the footsteps of these giant naturalists who, in some ways, gave up everything to dedicate their lives to this pursuit of wisdom. All of them suffered great hardships for decades to bring us the chance to learn what it is to be human and they did this by taking lessons from our closest living relatives.

Jane Goodall helped us learn one of the greatest lessons of all: that our similarities to both animals and each other are just as important as our differences. If it weren't for Dian Fossey, gorillas may already be extinct. As for Biruté? We've already seen the significance of her research. 'By being themselves, orangutans forced me to come to terms with my own human nature,' she says in her book *Reflections of Eden*[35]. Orangutans lead glacially slow, somewhat serene lives with little aggravation. Males live in isolation for years at a time and studying their life history is virtually impossible. 'In his eyes we see a precarious balance of ruthless strength and brutality on the one hand, and gentleness and security on the other,' Biruté writes.

Sowing division is a sign of weakness, of a breakdown of survival; for humanity to endure we must start seeing our differences as an opportunity, not a threat. Celebrating our differences by combining intelligences creates strength and this is how nature operates. For instance, learning to respect gender diversity within our own species is one step towards better cooperation with other species too.

To enable this better cooperation we need to turn to AI, or the 'actual intelligence' of a global network of animals that contains knowledge that artificial intelligence will never know about. Every day 20 quintillion lifeforms make autonomous choices that turn vegetation into habitat. Humanity depends far more on this collective intelligence (that we might call 'nature') than our own technology.

We should be wary of what the modern world considers intelligence. Instead of that word meaning something like AI, rather it refers to shared societal knowledge that comes from naturally evolved communication between members of our species. Artificial intelligence is exactly that – artificial. While it can be used to advance technical solutions under controlled conditions, it can also offer corrupt and misleading advice that draws on the opinions of a human-centric world that is disconnected from nature. It wouldn't be able to tell us that most vital lesson from nature that there is just as much cause to celebrate our similarities as our differences, because as a technology it is vastly distanced from any natural understanding of the world.

We must ensure we turn to the wisdom of people like Jane, Dian and Biruté who all have a lifetime's experience connected to nature, rather than a robot representing the 99.9 per cent of the human world who haven't got that knowledge.

The behaviour of animals and wisdom of people are the only mediums we have for realigning ourselves with nature. But that wisdom is an increasingly scarce resource that science and technology simply can't replace. Taking our lessons from nature by listening to our elders – our giants of conservation – and learning to understand wildlife is what will make us better humans. It's also the solution to some of our greatest challenges.

While Whales Rewild Us

As I was standing on the cliffs below Cape Nelson Lighthouse on the coast of Victoria I saw multiple whales breach beneath a double rainbow, creating an unfeasibly large splash that was visible for miles. A misty squall and light wind had flattened the sea as several humpback whale pods drifted by, breath from their blowholes visible as puffs of steam rising from the cold, blue ocean.

The size of whale pods depends on our perspective; we see groups of one or two animals along our coastline, but as whales grow in abundance they congregate and have even begun to feed cooperatively during migration. This is something living scientists had never seen until recently, which comes as no surprise to me given what little attention modern science pays to our natural world. In the 23 years I've lived here, I've seen the population of humpback whales increase from about 8,000 to around 35,000 animals. Everything has changed.

These whales have exquisite hearing and communicate using sound. They are aware of other whales around them up to distances of tens of kms, perhaps even hundreds[106]. It's barely possible for us to comprehend the scale of their communication, cooperation and impact on ecosystems and the benefits that brings to our farming, fisheries, weather, coastlines and so on. A society of whales – relatives and friends – know each other and share culture and knowledge like we do but in a different dimension. They could number hundreds or thousands of individuals travelling in unison, spread out over a vast area and connected by sound.

Like turtles in the Mediterranean and whale sharks in Indonesia, whale civilisations existed for thousands of years before we did. From the fragments of their culture broken apart by whaling, enough knowledge was thankfully kept alive by them to know how to start reconstructing their ecosystem. Whales are rebuilding the ocean's worth and doing this alongside all the other creatures that gravitate around their megalithic impact.

Just to the east of Bass Strait, where the Indian and Pacific oceans collide, you'll find animals that are traces of a once powerful ecosystem. While in recent years land-based recovery has been encouraged by people, the rewilding of the sea has developed thanks to intelligent whales. Whaling ended in 1978 and now humpback whales pass by unthreatened in huge numbers where they gorge on mouthfuls of fish and create their own ecosystem.

These whales are consuming and recycling nutrients and reintroducing traces of elements like iron, phosphorous and nitrogen into the water at concentrations millions of times greater than the surrounding sea. This enriches the sunlit surface and amplifies the initial land-based inputs to the point where fish populations can double. They are nature's guardians of the deep, fertilising and reconstructing an entire food chain from raw ingredients placed there by plants. This is natural rewilding on a giant scale.

> **LESSON 20**: Rewilding ourselves is a natural course for humanity. Whales are leading our way.

Although this process is happening in plain sight, there is no formal recognition of the value it brings to our local economy; the significance of these events and their impact on our lives remain invisible to science. We're still asking too much about 'how' and not enough about 'why'. Why does it matter that our whales have returned? It matters because this is the reason why thousands of years of civilisation settled here in the first place. The return of our whales is a clear and wonderful signal that nature has reached an exciting turning point where it is restoring an ecosystem without any input from humanity. It's marvellous, mysterious and every year that I see this regeneration of giant proportions it reinforces my hope for our future.

This overriding power of nature is what the Gunditjmara of Budj Bim and countless indigenous groups all over the world recognise. We should be celebrating this shared fate we have with wildlife because it is an essential lesson in survival through cooperation.

Connecting with wildlife isn't about buying an eco-conscious holiday, learning some fun facts about endangered creatures or going on a cruise with hundreds of other people to watch a parade of penguins.

It's about giving ourselves to understanding the power of wildlife and letting animals reconnect us to nature. By that I mean recognising that animals can choose to convene and cooperate with us on their own volition. When that happens it's a moment to be cherished, because at that point in time we know what it means to be alive. It offers a glimpse into our purpose and the collective role all animals play in shaping our planet and surviving together.

For the last 50 years we've increasingly mined our planet's fragile surface and presided over a decline in the local abundance of wildlife. We have made gigantic changes that have brought about all sorts of problems for our own and other species. We have not been respectful of nature and the delicately balanced ecosystems that we exist in.

We could continue to do this, or we could take a step back and leave nature's recovery in the care of the intelligence present in the brains of billions of animals. That combined intelligence of thousands of species acting together exceeds anything we can ever imagine, including our current obsession with AI.

We have only just begun to work out the vast benefits wildlife will bring our society, which is hardly surprising given that our egotistical and dogged belief in our own importance as a species has clouded our better judgement. Widespread adoption of nature-based solutions for environmental problems has only recently emerged in our social consciousness but uptake has been intense. The first global standards were published at the start of the United Nations Decade on Ecosystem Restoration in 2020[107], which sparked ambitious and exciting efforts by a number of countries to reframe the way we live. Finally, people were talking about us being part of nature – a momentous occasion.

Most of the world has signed up to the Global Biodiversity Framework for managing nature through to 2030[108], agreeing to protect one-third of the planet's most important ecosystems. The world has generally accepted that nature is now the key enabler to address all other existential crises, putting this one profoundly exciting new belief system into action. This change is happening because we need to meet economic challenges; our future is no longer guided be greed alone but instead is forged by the physics of living on a finite planet. This finality defines who we are and what our future will be.

We do know we have reached a crisis point and the only way we can transform society is by using lessons from nature to build a sustainable future. We are on our way to realising this new nature-based economy and this is hugely significant. It may seem like a small step for man, but it is a giant leap for humankind.

Guardians of the Wild

The Whale Sanctuary Project in California is trying to create an environment where cetaceans born in theme parks are returned to the sea, as one of the founding principles of this scheme is to give animals wellbeing and autonomy. This is a small step in the worldwide cry from conservationists to give animals rights of their own in line with humans and corporations. To understand why this action is significant, we first need to look at what is meant by 'autonomy'.

In the *Macquarie Dictionary*, autonomy has several definitions:

1. Government
     a) the condition of being autonomous; self-government or the right to self-government.
     b) a self-governing community.
2. Independence; self-sufficiency; self-regulation
3. *Philosophy* the doctrine that the individual human will contains its own principles and laws[109].

Only one of these definitions mentions humans directly. The rest are focused on the concept of leading one's life according to reasons, values or desires that are authentically one's own, as well as doing this in a community setting according to rules set by a learned society.

Animals have autonomy. We have seen the evidence in 'Civilised Green Turtles' that successful society is formed of communities governed by shared knowledge and communication. They regulate themselves. If by this human-created definition above they are autonomous, then they should (in theory) also have the right to self-govern without the need for our permission.

Unfortunately, in the eyes of most people the concept of autonomy has become synonymous with human knowledge. Almost all wild animals make instinctive choices where they stand, fly or swim based on the knowledge they received from their ancestors. They don't have access to the latest research, can't speak for themselves and know only these natural reflexes. This means their autonomy is easily taken away because it is assumed that they do not self-govern or self-regulate. In the eyes of mankind they have no rights because their intelligence takes a different form to ours.

Wise choices, those that are overall most compatible with functioning ecosystems, depend on freedom of movement and access to the right variety of options to choose from – what we might consider bodily autonomy. But all animals are acutely skilled at doing that when given the chance.

A crucial point is that, in our view, autonomy also depends on being *informed*. Though we might not think animals have this skill given that they can't read the latest issue of a scientific journal, they do and so do you. In fact, it is vital to our existence.

In nature, having up-to-date knowledge is the difference between finding food and not finding it, life and death and the collapse of a way of life or a whole civilisation. It's the same in human society. The danger for smaller and more fragmented wildlife populations is that they can't transfer knowledge so easily between members of their colony.

Pollution and deforestation of habitats puts life-supporting decisions out of date, leading to confusion and the eventual disintegration of the ecosystems they exist in. We've all seen images of orangutans or koalas climbing lone trees surrounded by forest that's been recently felled for palm oil or housing developments. No-one told them their knowledge had expired or that the landscape changed beyond recognition for them. Or the lonely one-horned rhino walking through the busy streets of Nepal to reach centuries-old grazing, still treading the path to ancestral grassland that has been known to its species for centuries. By doing all these things we are removing and undermining their right to autonomy.

Today there is nowhere untouched by humanity and, apart from scattered indigenous nations, one type of civilisation largely dominates the whole of the planet. We have become a contiguous people with no buffer between us except for the oceans, which are also being depleted. To believe that we can sustain ourselves in this current way of living is out of touch and dangerous.

When nature calls we respond and changes happen fast, inevitably for the better. But all too often we have a knee-jerk reaction that happens after we've reached tipping point on a particular issue. This occurs without full regard for what impacts that might also have on the other animals, so our efforts are often counter effective. As author Tyson Yunkaporta says in *Sand Talk*, 'Like all things that last, it must be a group effort aligned with the patterns of creation discerned from living within a specific landscape[77].'

To avoid panicked responses we need to start giving more autonomy to the natural world and we need to start now. But just like a child, wildlife cannot speak for itself in human laws and governments. This is where the idea of custodianship comes in. What we need is to be the guardians of our natural world.

In Isaac Goeckeritz's documentary *The Rights of Nature: A Global Movement*[110], he explains how Utah law still allocates about 80 per cent of its water for agriculture when that industry only represents about two per cent of the economy. Meanwhile, the Great Salt Lake of Utah is drying out. Water, as I'm sure I don't need to tell you, is one of the three key elements needed for the survival of all living things. Not only will the wildlife in this area rapidly disappear without this vital life source but the human population will, too.

Thankfully the people of Utah have realised this and are on the first stage of a new journey. They are trying to determine how to give rights to the lake to save it. One way to do this is to give it personhood. Giving personhood to landscapes or animals does not mean believing they are a person. All it means is that they are given equivalent rights to humans. As we have seen in earlier chapters, this process is already occurring elsewhere in the world where personhood has been given to rivers, lakes and even waves.

We offer similar rights to buildings under natural heritage protections, and even corporations are given personhood under certain laws. But when it comes to animals, this protective state of personhood is selective and only benefits the creatures that we determine are worthy of it. We have laws that allow us to protect our pets and you can't walk into a public zoo and take aim at the animals in the cages, but the wildlife around us has none of these protections.

Recognising the right of animals to exist and to have the benefits of all that autonomy means our economy can start to reap the rewards. Think of it like this: if you're a property tycoon who wants to build hundreds of new homes near an ecologically sensitive area, would you prefer contributing seven dollars more to the local economy for every dollar you spend building? Or spend millions of dollars over many years and risk having your development rejected? The first option means your properties are more valuable, as banks will invest in your development and insurance will cost you less. The decision-making authority can ensure it is socially and ecologically sustainable, meaning you could get up to get eight times the economic return with the bonus of voters that love you. Seems like a no-brainer to me.

Some might say wildlife already has autonomy through our existing legal system but that is paper thin and drawn from the ledgers of legal drafts made decades before we were wiser. In fact, most local laws still prohibit us from standing up for the basic rights of the ecosystems we live in.

South Australia's *ERDC Parliamentary Inquiry into the Urban Forest* in 2023 found an estimated 11 per cent canopy cover loss in a decade, equivalent to about 75,000 trees a year. Urban heat modelling shows that parts of Sydney could fail to provide habitable shelter in a few decades. Street trees reduce temperatures by up to 25 °C and if their numbers decrease it will affect our day-to-day lives. Meanwhile, the rights that should be given to nature are instead being given to property developers. Primary koala habitat corridors are being destroyed every week to make way for construction and development, even though koalas are essential to forest and human health.

It's already common for governments to use corporate personhood to make decisions about nature. For example, businesses can be given permission to destroy nature if they pay a premium and promise to restore it elsewhere. As we've seen in previous chapters, human intervention in attempting to recreate an ecosystem that existed elsewhere often upsets the balance of natural life. These animals and habitats have developed in a certain area for a reason and modifying them in ways that reduce nature's ability to work for us (or shifting them elsewhere) are poor solutions because they are simply not meant to be that way. Additionally these offsets for businesses are corruptible and can be gamed to create political outcomes.

Offsets only work if *irreplaceable* biodiversity is protected. There is little public land left that is replaceable, so governments have tended to allow it to be destroyed. In one example, the Victorian Government in Australia offset fragile, endangered grassland in favour of development but couldn't demonstrate that habitat's capacity to recreate something of equivalent value elsewhere. Why? Because replacing irreplaceable ecosystems is an oxymoron.

To survive, communities need guaranteed clean air, unpolluted water, un-poisoned food, abundant wildlife and affordable homes. What they don't need are industrial-sized developments that destroy nature. But when it comes to the fight for the rights of wildlife, most wins tend to sap the time and resources of conservation groups. As Judith Hoyle (the campaign head for the Toondah Alliance who we met in 'High Flying Achievers') laments, 'It became a full-time job.'

Absurd conflicts create anxiety and instil distrust in decision-making authorities. Corporate leaders know this leads to discord. It's the type of behaviour that results in uncertainty for investors, which is why giving businesses custodianship and rights to determine the lives of wildlife can also be detrimental to the economy. Take the case of the 152 koalas euthanised and 79 sterilised on behalf of a US aluminium company Alcoa in Portland, Australia. The cull sanctioned by the Victorian Government and Zoos Victoria gave the main reason as being that the animals were suffering as a result of 'overpopulation' and were in pain[111].

Overpopulation is a made-up term that has no place in ecology. Populations increase in response to an ecosystem changing and it's not the animals who are to blame for this but rather the threats that broke the balance of their ecosystem. Overpopulation is a convenient catch-all term used by people who don't want to accept that the inconvenience of having more animals was caused by us.

What is actually the case is that due to the outside pressure of land degradation the koalas had been forced into the only areas where suitable forest remains – a location that happens to be near Alcoa's smelter. Locals say koalas are killing the last remaining manna gum forests, which is true, but they have no other option because we've removed their habitat without rebuilding it elsewhere. As koalas are forced into smaller and smaller areas this has created conflict with humans, rather than recognition of the opportunity for peace that koala rewilding could bring.

Meanwhile airborne pollution from Alcoa's aluminium plant has led to crippling bone disease for the marsupials that live there. Somewhat surprisingly the Environment Protection Authority has given that pollution the all clear for people who live and work nearby[112]. Worryingly, it's exceedingly common for acute diseases to be found in mammals and ignored in humans, and until we accept the similarities between people and other animals we will continue to put our own society in jeopardy.

It may also surprise you to learn that custodianship of remaining koalas has also been given to Alcoa. This 'get out of jail free card' takes the form of an ongoing management plan and responsibility for the lives of another 120 of these vulnerable creatures. In a perverse twist, a koala that was previously dispensable to Alcoa could now be considered a corporate asset. By permitting Alcoa to continue and by giving them control of the lives of individual animals, the government may have inadvertently put a value on each koala by ensuring they could be added as assets to Alcoa's balance sheet. It has used a form of personhood to protect them. On the plus side, Alcoa is now saddled with that responsibility, so in some respects it's less of an asset and more of a liability.

Our incumbent laws have served our economy well until now. Increasingly they are out of date and out of touch with the state of the planet. If koala management had been done through a nature fund and run independently, Alcoa could have claimed this as an investment and the government may have been able to unlock other corporate finance as part of that negotiation.

If the koalas were protected by people who care more about them than about profit, this could unlock eight times more economic advantage for the broader community[41]. Why? Because animals unlock the benefits of nature recovery and this would ultimately make all local businesses more sustainable. CEO of Swiss Re, Christian Mumenthaler, makes the point that 'today, 55 per cent of global GDP is moderately or highly dependent on biodiversity and ecosystem services'.

This is because businesses and communities aren't separate entities – they all depend on each other and on a healthy ecosystem. The best way to look after nature is to give custodianship to those who already believe animals have an inalienable right to autonomy and who can work on landscape-scale endeavours to restore the balance of nature. Rather than money being thrown at a clueless corporation, this guardianship ensures that cash flows back into rebuilding broken communities and is a long-lasting solution.

Corporations are inherently focused on their own profit and can be commended for supporting the livelihoods of their own workforce. However, it's generally accepted that on average investment in nature-degrading activities only creates a dollar in value back to the community. The government backs business because it's the only way they know how to generate this dollar and this marketplace is what has enabled corporate personhood to evolve.

Everything was fine while we were reaping the rewards from business and basking in the free services that ecosystems provided (that we took for granted). As those natural services have disappeared so, too, has most of the value we were getting.

What we now know is that every dollar spent on restoring nature can recover over 30 times more value back to the community. To unlock this potential that benefits both people and businesses requires a different approach to using personhood.

This is where Rebalance Earth and Blue Green Future are daring to tread new ground. Having worked for decades to help leaders stabilise their economies through good governance, it was a natural step for Ralph Chami and his colleagues to turn their minds to transforming the marketplace for the new nature-based economy.

By 2022, some Māori tribes in New Zealand had become aware of Ralph's work at Blue Green Future and they met at the COP27 climate change conference in December of that year. The Māori were already worried about change in ocean health, the impact on whales and how this would affect their food security. By June 2023 they'd launched the

Hinemoana Halo ocean initiative with Mere Takoko, vice president of Conservation International, who had brought together tribes from New Zealand, the Cook Islands and Tonga. All parties signed an agreement to collaborate on various nature-based projects and Blue Green Future were asked to help shape those projects so they could go to market to get investment for them.

For nature to have a valuation and be taken seriously by the market economy it can either be deemed as an asset, a legal person or both. There was a precedent for this already in New Zealand; in 2017 the government had given legal personhood to the Whanganui River and bestowed custodianship of it to the local Māori tribes. Māori consider whales to be their ancestors, so the natural first step was to seek declaration to confer personhood on the whales. This was supported by the Māori king in the Cook Islands on 27 March 2024.

This is only the start of a process of recognising the autonomy of animals that will ultimately create one of the largest protected areas on Earth driven by local people, enabled by government and financed privately. It doesn't matter where a whale is killed, if it is recognised in law as having the value of personhood and there are human custodians, there becomes an imperative to protect it.

Referring to a cruise ship striking and killing a whale, Ralph says,

> Markets simply see no value in protecting the whale, and this ship is an example of the markets' ambivalence … Conferring legal personhood on the whale, as the Māori people did on March 27, is a game changer. A living whale is now 'visible' to the market. The living whale has now a right to free living and pursuit of happiness.

We must be careful that the implementation of these protections doesn't become a path for larger bodies to lay blame at the feet of individuals and employees. We are all victims of injustice and circumstance and wise people know this. A ship's captain must meet deadlines. He works for a company and society that doesn't value the whale and so his decision to not avoid a whale in the path of his ship isn't a decision he's made for himself. Instead, it is one made for you and me due to an economy and collective mindset that renders the whale insignificant. It isn't his fault any more than it's your fault your local primary school stocks plastic forks in the canteen. However, give people the opportunity and everything changes. They learn to care.

Ralph wasn't simply able to work out the value of whales that are bestowed personhood, he could also determine that the services they produce could be made into an investable commodity. Over the last five years this has begun to transform the way we make living, rather than dead, animals visible.

As one of the first projects of its kind, Hinemoana Halo has raised millions in rewilding bonds and the venture could eventually be extended to every island nation on Earth. A protected area of 2,200,000 km² will grow to include most of the Pacific, Atlantic and Southern Ocean. The local people and communities that embrace this action will soon be able to invest in animals other than whales, building connections between land and sea and enabling the restoration of entire coastal economies, culture and community.

Rather than trying to patch up or retrofit failing economic models, what's happening is the redesign of whole economies by incorporating into business the concept of supporting nature as a benefit. Three hundred and twenty of the world's largest companies have just begun reporting nature deficits in their annual reports. The Taskforce on Nature-related Financial Disclosures (TNFD) is to become a legal requirement for many banks and insurers. They are understandably nervous because they may have lent money to businesses that haven't considered all the risks. Their loans may be at risk of default unless they protect nature.

Investments that consider nature will be less risky, as well as helping entire nations to restore their culture, dignity and future by once again riding on the shoulders of giants. All it takes is to give animals their autonomy by granting them the right to exist through bestowing guardianship of them onto the communities that live among them – the communities that have the most to lose if we lose wildlife.

Letting Go Control

The idea that humans can manage wildlife is a given if collective (perhaps egocentric) perceptions of the world are to be believed. Humanity loves a hierarchy and we are determined that animals are beneath us and in need of our guidance. However, there is overwhelming evidence that such actions undermine our chance to restore ecosystems and, as we saw in the previous chapter, there is a now a need for humans to recognise that they must be guardians – not managers – of the natural world.

The reasoning behind this is simple; animals do a much better job of managing themselves and their habitats than we ever could. We presume knowledge of them and their cultures without ever really stopping to find out whether what we think we know is the truth.

When we remain closed off to the idea that wildlife creates healing processes when left to itself, we end up picking at the wounds we've made by never quite allowing things to recover to full strength and vitality. In addition to that, there is the shifting perception of what we consider 'normal' and how much ecology has often become viewed through the lens of nostalgia. The urge to return things to how we remember they once looked – rather than what they should be – can mean we build barriers to nature's restorative progress. To rebuild a habitable world we need restoration by animals, for animals.

Rewilding by nature's needs is our simplest, cheapest and quickest course of action towards a more habitable planet and the best way to secure our survival for the next 100 years, but to reveal that magic means putting aside our past methods of doing things and relinquishing control.

Why should we do this? In his book *Rewilding the Sea: How to Save Our Oceans*, Charles Clover asks one simple question: 'What is the natural state of the sea?'[113]. He comments on how there isn't much wilderness left and having depleted so much, surely we must restore it full of fish and other large animal life? But how can we know the true state of our vast oceans given that we dwell on land? Therefore, how can we rewild it with populations of various fish and other animals if we don't have this knowledge of what should be there?

Rewilding means different things in different places. In New Zealand it mostly means eradicating pests and reintroducing native species. In the US it's been about allowing native wildlife to rebuild itself in wilderness areas. In the UK and Europe it's more about finding a way to use species, including traditional domestic breeds, to fulfill the role of long-extinct native wildlife. In fact, it doesn't matter how rewilding is defined because the principles are the same. It comes down to four simple concepts:

- Remove the threats.
- Give absentee wildlife a helping hand to reintroduce itself.
- Stand back and let nature take its course.
- Work out how to build a new economy while learning to live with nature.

The concept of rewilding hasn't gained its full reputation among scientists yet because it is not possible to study how it works until it happens. Often we're distracted by discussions such as whether wolves in Yellowstone National Park really do benefit the environment, only to find out that scientists can't live long enough to prove it anyway. Charles talks about how 'science advances one funeral at a time as the deaths of fierce advocates of past orthodoxy allow new ideas to gain ground'. By removing our outdated perspectives of what should be done, we leave room for newer and more effective methods to be put in place. You're doing this right now, just by reading this book.

However, in some cases that leaves us at the mercy of everyday people willing to challenge the orthodoxy with their own money, ideas, inclination and a vision to rewild the land. This can be positive in raising awareness and beginning to action changes to our ecosystems but also has a negative side in that rarely do people act without consideration of their own agenda.

Therefore, it is fundamental to consider the order in which we approach these ideas. There are three interrelated components needed for successful nature-based projects to excel. These begin not with scientists or politicians but with you. We need:

- A belief in the right of nature to exist.

- The knowledge to understand ecological risks.
- The autonomy (for both people and animals) to make socially acceptable and ecologically relevant decisions.

It is only through these factors that we can begin to work to ensure the survival of humanity in future.

> **LESSON 22:** If we stop meddling and enable nature to take its course, we will rapidly strengthen our economy.

In Eastern Indonesia the Kei Islands are among the poorest places on Earth. The annual income for villages there can often be no more than a couple of thousand US dollars. In 2017 the OceanEarth Foundation began a project called SeaNet Indonesia to help those local communities[114]. The project was designed and led by founder Anissa Lawrence, a reformed accountant and conservationist.

Rather than telling the community what they should be doing, she asked, 'What do you need most?' The answers were what you might expect with income for their children's health and education being among them.

In one village the men would go to sea fishing each day while women stayed behind. On return the fish were cleaned and significant protein discarded. As part of SeaNet, villagers sanctioned the creation of new supply chains where the women would process this previously discarded protein into new high-value products. Agreements were made to sell these to markets in Ambon, the biggest city in eastern Indonesia. The women were trained in marketing and guided in how to get food-handling permits.

Meanwhile the men's traditional fishing knowledge was applied to another problem; during fishing trips, bycatch of turtles and dugongs would often result in damage to valuable nets. It turned out that rebuilding larger-gauge nets solved this while also increasing the number of target fish caught. Over a few months the community was able to double its income and almost completely eradicate the accidental catch of non-target fish, including the all-important reef engineering species and other marine wildlife such as turtles. A community knowledge-sharing centre was built to transfer learnings to other places, and seven years later the projects are still running without the need for outside funding.

The most valuable lesson from this is exemplified in the following event. A manta ray had become trapped in nets and the villagers proudly saved it. Co-designing, co-managing and co-implementing this project with local people hadn't just saved endangered species, it had also sparked the unifying human desire to act to protect wildlife. It brought the mindset of conservation and sustainability to the forefront of the minds of some of the poorest people on the planet, and by giving them purpose, connection, hope and income it revived that part of their culture that is inherently connected to nature.

When it comes to conservation Anissa says, 'There is a need for integrated conservation and poverty-alleviation style approaches. Nature-based solutions offer the ability to do that because we're addressing the priorities of societies rather than what we think is best for people.' At the same time as SeaNet, Anissa's team were working on a bigger project called GhostNets Australia instigated by Traditional Owners in Australia's Gulf of Carpentaria.

Ghost nets are nets lost at sea that drift around or settle on the seabed, entangling wildlife. These Traditional Owners are the custodians for over 3,000 km of coastline where up to 9,000 tonnes of abandoned, lost and discarded fishing nets wash up every year. The nets are making the country sick, and by healing the country the Aboriginal and Torres Strait Islander population believes we can heal Australia's people too.

After the discovery that about one-quarter of the nets in Australia were coming from Indonesia, these two programs merged to become one of the most gloriously ambitious and thoughtful conservation projects this century – the kind of project that will innovate and inspire great change, sequels and spin-offs.

The reality, of course, is that it takes time to grind away the inefficiencies and clean up a system that – like the exhausted whales towing fishing nets entangled around their tails – is shackled to the inefficiencies of our past. Each component must be unravelled piece by piece before our economy can be rebuilt with renewed resilience. Once we have removed these mismanaged influences of human behaviour, we have enabled nature to take its course in repairing any damage that's been done to the ecosystem in the meantime.

Why should we accept that this will be any different to the failed attempts at offsets or carbon credits in the past? It's a good question.

Nature capital accounting is different from investment or philanthropy. It's about seeing natural capital as an asset that can be invested into. It's a third option where credits in biodiversity can increase in value independently. It doesn't matter all that much whether the 'credit' relates to carbon, biodiversity, marine plastic or something else. We don't invest in one form over another. We make sure that all forms of investment allow nature to recover as a means to reduce the threat to human and all other animal life.

Let's take an example. The Arafura Sea has a lot of seagrass and animals that graze on seagrass, including turtles and dugongs. Nets impact the sea floor and entangled turtles and dugongs significantly reduce the amount of carbon being captured. As we will discuss further in 'The Fear Effect', grazing animals (especially in concert with predators) enable up to 80 per cent more carbon to be converted into plant material each year.

It is possible to put a dollar figure on that process and sell the retention of that carbon. However, the entire basis of that 'investment' hinges on ensuring the outcome will be long lasting. Unless there is unadulterated commitment to addressing the cause of seagrass decline, we would be unwise to hedge our investment. This is why a commitment by the community as well as proper governance is essential; in other words, the viability of any investment in nature and our future is influenced by outside pressures. Our economies are about to become dependent on making sure our belief in nature's value is also coded into laws and business practice. That's never happened before.

Let's go back to the rewilding concepts we mentioned earlier. Our minds tend to work forward through problems, starting with threat removal. Traditionally, this is why we have jumped to putting fences around protected areas. We now know this never works long-term because it excludes local people who are likely custodians of the wildlife in that area. Thankfully this realisation is dawning on us and is one of the reasons why national parks that have previously excluded indigenous people (like the Kattunayakan elephant custodians in India or the Māori whale guardians in New Zealand) are beginning to regift custodianship to those communities.

The main thing we must do is the last thing on the minds of most scientists; OceanEarth Foundation's GhostNets and SeaNet projects show us how successful we can be when conservationists first work with communities to build new policies that embrace nature-based solutions. The idea that we should not manage nature but step back and allow it to take its course is one of the biggest global economic shakeups to have ever happened. In the face of a crisis, and because it's both cheaper and faster to act this way, it will intuitively become the dominant human behaviour. If nothing else, this is a source of hope.

What's even more exciting is that it's being led by young people. 'The generations coming through behind us have raised the marine waste issue to the point it now has global provenance,' says Anissa. 'We even have a global treaty that is being developed as we speak.'

In the same way as our bodies are open systems connected to everything outside of us, nature-based solutions also need to sit within a supporting system. If we influence these systems with purely man-made or human-led processes, then the success rate will not be as high.

The big question is, why should I care? Why should the average Australian, or indeed anyone, worry about ghost nets and dead turtles in the remotest part of the Gulf of Carpentaria, or the poorest villagers on Earth in the eastern Arafura Sea? It's simple, really.

Relatively few people live where the most important life support systems are left on Earth. Here people such as the First Nations rangers in the Gulf of Carpentaria off Australia's northern coast are doing a huge service for humanity but they must constantly do beach clean-ups. It's the kind of activity that Professor Richard Banati of Australia's Nuclear Science and Technology Organisation once referred to as 'glorified garbage collection'. This is the crux of the issue: so much conservation work ends up pouring finance into cleaning up a problem rather than addressing the cause. The more it keeps happening, the more we clean up. It's a financial black hole.

When nature is invisible to our everyday financial needs we struggle to care about it. Now it's becoming visible and relevant to our everyday lives, affecting everything from regular rainfall to food security, climate warming and everything in between. Just consider how much economic loss we suffer when communities responsible for the health of our coastlines must direct their work to beach clean-ups? Instead, what if they could be growing turtle and dugong populations and allowing nature to thrive unimpeded by human intervention?

By now it should be clear that enabling nature to take back control is the fastest way to unlock the economic potential we have and address the rising cost of living while securing our long-term future. This is already happening naturally to some degree, but there are also important changes occurring in our own society. The benefits of nature-based solutions are to support business while also maintaining the life support systems that maintain our livelihoods and lifestyles. But to do this means unfolding new measures that allow communities to co-design the future they want. I would never have heard policymakers talk this way 10 years ago but now it's commonplace. The desire and need to do this is even outpacing our readiness to put this into practice. That's a good thing, though, because it means the foundation is solid.

In my hometown of Melbourne, community-designed coastal planning policies are being created so quickly that most responsible authorities aren't even aware of them. Soon, decisions may need to be co-designed and co-managed by the community but no-one really knows how this will play out. All we do know is that governments can't continue to centrally manage everything and without letting go control they will grossly inhibit their chances of receiving nature-capital funds. Protecting lives and livelihoods in a city that was voted most liveable city on Earth for seven years running is politically, socially and ecologically expedient.

Those who fund nature-based projects must see them led by local and indigenous wisdom, as well as these initiatives being supported by policy that protects nature in perpetuity. They must assign people to act as guardians, not managers, of our valuable natural world.

I agree with Goodall when she says the important role for scientists is in supporting societies to rebuild the habitable planet they dream of by stepping back and allowing these processes to be led by nature. We should not be taking over and dictating what should be done.

As Dinah Nieburg from Blue Green Future reminds us, these changes are happening for a reason:

It can feel really devastating, but I can see that everything is sprouting. It's like winter to spring. It feels like we're at a turning point and we need this transformation. The systems that have worked in the old world aren't going to make it anymore. It's time to show each other another opportunity.

We really should be living on a planet where we don't have to be constantly cleaning up after ourselves or managing nature just to stop it falling apart at the seams. That world where we can live without constant worry is within our grasp, but whether you believe it or not, enabling nature means doing less not more.

It is time to sit back and go with the natural flow.

The Fear Effect and Food

The world was to me a secret which I desired to divine.
Curiosity, earnest research to learn the hidden laws of nature,
gladness akin to rapture, as they were unfolded to me.
FRANKENSTEIN, MARY SHELLEY

It was late afternoon when we crossed the East Alligator River in Australia's Northern Territory. As our vehicle divided the water on the causeway, a few crocodiles slipped beneath the surface. The river marks the northern boundary of Kakadu National Park and the gateway to Arnhem Land and the remote town of Oenpelli. The anticipation was palpable for the journey ahead as we embarked on the final stage of our pilgrimage to a land of grasshoppers that are imbued with the power of lightening and storms.

As our tyres hit the red dirt road we emerged from the riverbank to a vista that took our breath away. The distant horizon was marked by a line of cliffs glowing vermilion in the warmth of the afternoon sun. Well into the distance the river meandered over a floodplain edged with vibrant green reeds and smoky-grey eucalypt forests. A lifetime on any continent can't prepare you for the feeling you get when you finally set eyes on living landscape such as this where iconic animals still outnumber people. We were still two hours from Awunbarna where we were to meet leaseholder Max Davidson.

Max was invited to Arnhem Land in the 1970s by his friend Charlie Mungulda, one of the Traditional Owners of that land. Charlie was the last native speaker of the Amurdak language and helped approve the lease that led to Max setting up the lodge we were headed to. The money generated by the venture has been poured back into protecting local livelihoods, landscapes and culture. Tourists come here to see 50,000-year-old rock art, including an eight-foot-long rainbow serpent Max stumbled across in 1987. It's considered one of the most spectacular pieces of ancient art in the world.

But what we'd really come to see was Max's proudest legacy: the unlikely union between Max, an ex-farmer, and the endemic Leichhardt's grasshopper. Local Kundjeyhmi speakers call them *alyurr*, meaning 'children of the lightning man'. According to custom these grasshoppers emerge to call out to their father Namarrkon who sends powerful storms. Max was besotted with this creature, knowing from the wisdom he'd gained while spending time with the First Nations people of Arnhem Land that it would be key to revitalising their bushland.

Alyurr feed on native foxglove or *Pityrodia* that grow best after cool season burns. Their larvae emerge to spend their whole lives on a single bush. By the time they metamorphose into adults they are about five and a half centimetres long with a beautiful flame-orange body, black-speckled wings and glittering deep-blue backs.

Author of *True Tales of an Outback Guide* Mike Keighley, says 'People who live in cities and who've learned to survive in that environment look *at* landscapes ... Aboriginals look *into* them'[115]. Keighley's co-author, journalist Tom Huth who started his career at the *Washington Post* during Watergate, once described Mike as 'the real Mick Dundee'.

(On a slight aside, Kim Kardashian once tried to hire Mike as a guide but he declined. He was leading a school group at the time, who scarcely believed it was her people he'd been on the phone to.)

Through a life on the land, lessons from wildlife and understanding ancient culture, people like Mike and Max have had the good fortune to learn to look into the bush and come to understand their place there. Max was able helped rewild one of the most beautiful places on Earth long before the term was even invented. On that trip as we walked with Max over ground scorched by carefully controlled cold burns, he reached his hand down to touch the leaf of a plant emerging from the red dirt.

'What the Leichhardt's grasshopper shows to me is what can happen if you want to preserve species,' he told us. 'We have taken an animal from virtually on the edge of extinction. We look back and we see now that our whole bushland has changed'[116]. His work was only the start of understanding this relationship between man and this storm-summoning insect. More recently we've started to learn other extraordinary lessons from grasshopper rewilding; it's revolutionising how we think about the connection between climate, carbon, soil, wildlife, stress and major health crises facing humanity today. This is a story that captures the essence of almost everything we've explored in this book so far: ecosystem engineering, health, climate and our survival as a species.

Professor Oswald Schmitz of Yale University wasn't sure he was cut out to be a grasshopper biologist at first. 'I wanted to be able to study the rules of life,' he says. At the time, funding wasn't available to study large herbivores so Oswald decided to start his investigations on a smaller level.

What he discovered in studying the grazing grasshopper community is that there are layers of body sizes that act in parallel to the community of herbivores we're more used to observing such as bison and deer. 'Grazing mammals don't compete strongly with each other,' he says. 'They compete strongly with their likeness in the insect world.' Australia's biggest grazing megafauna died out 40,000 years ago. Is it possible that grasshoppers took over their role?

'Grasshoppers are not just automatons,' Oswald says. 'They have cognition, they are sentient. Like mammals, they're making choices based on rationality. It's not instinct because they're confronting new situations every day.'

The food plant of the Leichhardt's grasshopper, native foxglove, only thrives where its bushes spring from fires caused by lightning storms. Here moisture is held in the deepest soil layers that are packed full of organic carbon. It stands to reason that the grasshoppers must be making decisions that ultimately have an impact on soil carbon and therefore climate. They would have been constructing and maintaining this ecosystem alongside kangaroos and people for millennia.

Until recently, scientists hadn't thought to ask how the behaviour of animals on land might impact the food web, likely because it's hard to prove without modifying an ecosystem. Oswald's unwilling shift into studying micro-herbivores meant he and his colleagues could experiment by taking the system apart on a smaller scale and rebuilding it to see what happened.

At first they created a grassland with no predators, then they reintroduced the most common spiders. The results were illuminating. Introducing ambush predators such as web-building 'sit-and-wait' spiders into the grassland had no impact on grasshopper numbers but a huge effect on the richness of the vegetation. Then when active hunting predators like jumping spiders and wolf spiders were introduced, the dynamic changed again. This time grasshopper populations declined but vegetation health and carbon uptake increased massively.

Both responses were being driven by a different type of fear effect, and the difference between the results Oswald saw lies in understanding the two types of stress. Fear of predation is an acute stress and if it happens occasionally it's a natural response that improves a creature's health. This is so-called 'good stress' and it improves cognitive function because a bit of a shock makes your body create proteins that build new nerve fibres in your brain. It also protects us against infection and even enhances childhood development.

Chronic stress is different and reduces health as it's borne from a continual simmering worry and leads to despair. It was found that the result of losing control to stress was similar for insects and all other animals.

With active predators removed, grasshoppers could move more freely and choose to forage in the richest habitat. The grasshoppers would gravitate on mass to places where plants contained more nitrogen (protein). However, this would also attract more web-building 'sit-and-wait' spiders, so the grasshoppers were constantly on edge because they knew the spiders were lurking close by and could pounce any time. This elevated their fear response to something akin to chronic stress. When any animal becomes chronically stressed it craves more sugar and becomes obese. The grasshoppers then began to select plants higher in carbohydrate (sugar).

By now you're almost certainly thinking this sounds like work-stressed humans dining out on fast food. It's exactly like that. The fear effect modifies our will to consume different nutrients. Authors of *Eat Like the Animals*, David Raubenheimer and Stephen Simpson of the University of Sydney, proved that animal nutrition is regulated by our access to protein[117]. Too much and we die younger but breed faster; not enough and we become obese, stop reproducing and live longer.

When the authors first presented this finding they were shunned simply because it flew in the face of conventional theory. How could a couple of insect ecologists predict what nutritionists had missed? It took a few more years for people to come round to the idea but, ultimately, studying grasshoppers would revolutionise our understanding of human health and nutrition.

When all else is equal (and we have adequate autonomy) all animals are equipped with a range of inbuilt senses that enable us to make optimal choices. At normal levels of stress these senses help us limit our protein consumption, which increases the rate at which we can breed. This is a stable evolutionary strategy given that breeding is the reason why any population can survive.

But at times when protein is scarce and our stress levels are heightened, we consume more carbohydrate so we can live off our body fat. We don't consume more carbohydrate because we need more sugar, it's because we need more protein. But we must eat larger quantities of available sugary foods to meet our overall protein budget. This is the protein leverage hypothesis that Raubenheimer and Simpson discovered. This buys animals time to withstand periods of famine and use their mobility superpower to find new places to eat.

All animals, including humans, are exquisitely adapted to make these choices and it's been proved categorically. It's another reason we're not so different to all the other animals on Earth. The same hormone systems are present in the cells of insects as in humans and other animals.

Heightened stress alters our natural balance. Being overworked, consuming coffee or living with toxoplasmosis can all lead to chronic stress, increasing our metabolism and making us crave sugary foods. Chronic stress can become a contagion that leads to an individual's heightened risk of diabetes, obesity, autoimmune and other diseases[118].

Consequences such as diabetes and heart disease might only affect some of us over a long period of time but this makes the population-level effects more sinister. A population that becomes too chronically stressed overall will decline because obesity leads to lower reproduction. We have little control over this. The American Psychological Association even defines sources of chronic stress as things that will 'change people's identities or social roles [and] are beyond their control'[119].

But is it beyond control? Ultimately, stress originates from a breakdown of the environment around us. What if the solution to freeing ourselves from this fear effect lies in rewilding the landscape? After all, our behaviours are triggered from the outside in.

There is evidence that this could be the case with other animals. Animal physiologist Edward Narayan of Stress Lab at the University of Queensland has shown that the level of stress hormone in koalas living with land clearance is 15 times higher than the average stress hormone levels in koalas living in plentiful habitats. Habitat loss is now thought to be the main reason why koalas are suffering from diseases like chlamydia. Stress might also explain why some fragmented koala populations end up doing more damage to remaining trees, because once there are few trees left, chronically stressed eucalypt specialists like koalas would seek remaining foliage with the sugariest canopy.

In times of extreme ecosystem pressure the 'fight' response might be substituted for 'flight'. Mass exoduses of animals searching for food with more carbohydrate means animals get together to stock up on body fat, survive periods of protein scarcity and find mates. Nutrient imbalance due to overfishing could also help explain why shark densities seem to be increasing in some places despite world shark populations plummeting. Like animals gathering at a watering hole during drought, these congregations are essential to survival.

Humans, meanwhile, have given increasingly stressed people convenient access to sugary foods. Simultaneously we've introduced all manner of primers to further increase stressful behaviour and that is not natural. It's perhaps no accident that coffee – one of the most valuable commodities on Earth – is responsible for massive ecosystem destruction. This has made us more prone to diseases including those coming from wild animals. The question is, what can we do to fix this?

Cast your mind back to the second step in Oswald's grasshopper experiment where he reintroduced active spider predators into the grassland. This time the grasshoppers had to alter their behaviour to avoid the occasional risk of being caught by the predators when they wandered close by. This type of stress isn't dangerous as long as it's only occasional. Having wolf and jumping spiders around instilled enough fear in grasshoppers to limit their foraging distance. They become more thinly distributed and, because they were minimising the risk of capture, the overall population ended up eating a better balance of carbohydrate and protein.

This is where the connection with carbon and climate comes in. Protein contains nitrogen, the raw ingredient of most fertiliser. It is consumed by soil microbes and used to nourish vegetation. Carbohydrate-rich foods contain much less nitrogen. The fat, chronically stressed grasshoppers over ate the sugary plants meaning there were two consequences when they died. First, their obese bodies overwhelmed soil microbes by depositing too much carbon. Secondly, because this only happened in certain places and contained little nitrogen, they deprived the general environment of nutrition. This led to reduced plant diversity, less carbon-rich vegetation and poorer soils.

After active spider predators were reintroduced, however, the less abundant protein-eating grasshoppers' bodies became leaner and richer in nitrogen. The fear effect stopped them travelling further but chronic stress declined as there were fewer ambush predators. When these animals died, even though they weren't as fat they were better for the soil microbes in the environment because they provided it with more nutrition.

A mere five per cent change in the balance of carbon to nitrogen in the carcasses of grasshoppers radically alters the amount of carbon reaching the soil. Oswald found active spider predators were responsible for a staggering 40 per cent of soil carbon processes. Imagine that on a larger scale and it's soon clear that whether it's wolves or wolf spiders, the presence of active predators rebalances our ecosystems.

I'm not advocating introducing something like wolves into our suburbs. What I am suggesting is that rewilding our landscapes overall with predators or otherwise could reduce stress in *all* animals and help tip the balance back in favour of health, happiness and a more stable climate. Based on this possibility it's not much of a stretch to consider that stress in humans could be both caused by and the cause of more disconnection to nature.

A friend of mine, Peter Lamshed, gives counsel to people suffering appalling trauma. 'Despair,' he says, 'comes from a lack of information and feeling like you have no control over your own life.' Peter, who is deeply passionate about wildlife, also knows that trauma stems from a disconnection from our natural environment. It's a case of becoming 'disconnected from our "feeling state",' he says. 'We even call it "the" environment when perhaps we should call it "our" environment.'

As the author of *Wilding: The Return of Nature to a British Farm*, Isabella Tree, says, the need and ability to connect with nature on a practical and even spiritual level is 'in our genes. Sever that connection and we are floating in a world where our deepest sense of ourselves is lost'.

I'm sorry to be the one to tell you, but a healthy reconnection with nature means breaking our addiction to coffee, work and 24-hour news. The thing is, you already know that. Doctors all over the world have started issuing 'green prescriptions' knowing that as little as a couple of hours a week in green space can have significant health benefits. Though I'm doubtful that walking in a park full of trees is the solution. The gold standard is rewilding your neighbourhood, connecting with wildlife, birdsong and embracing the smells, tastes, textures and diversity of a functioning ecosystem.

Stress doesn't just cause problems on a personal level but extends out to a societal and even global scale. When we degrade ecosystem services our taxes and cost of living increase, meaning we must work harder and longer. The subsequent despair is real. The resulting stress makes us crave carbohydrates and our society capitalises on this by processing more sugary food to fuel a growing addiction. We are eating ourselves out of house and home.

Climate change is increasing the concentration of atmospheric CO_2, which also leads to crops with less protein. That and the fact that three-quarters of our diet consists of low-fibre, low-protein, ultra-processed foods that speed up absorption of protein and result in the need to eat more. Ultra-processed foods also don't contain enough protein so they are killing us slowly. We have tricked our bodies into thinking we must eat larger and larger portions of carbohydrate in the desperate pursuit of a protein fix that's never quite achieved.

You might say humans are becoming a plague and that would be an accurate analogy. Plagues form when animals are forced to consume more carbohydrate. In locusts this happens when they are deprived of protein. Their metabolism changes and they transform into flying fat-stores, intent on drawing out their survival for as long as possible by eating as much sugar as possible until they can once again find a balanced source of sustenance. Humans have more than enough protein if we want it, but we have built an artificially unintelligent new food system that increases our stress and leads to our sugary craving. We've commercialised the most addictive properties of food that we crave most when we're stressed and created for ourselves a more stressful environment to live in. It's one built for profit and makes us less healthy because we're disconnected from nature.

Fuelling our own addiction to carbohydrates might mean modern farming has created the perfect storm. Plagues of mice and locusts happen more frequently where predators have been exterminated and fertilisers are used to encourage the sweetest fast-growing grasses, weeds and crops that are high in carbohydrate and low in protein.

For Oswald, these small animals are what he calls 'analogues' for the larger animals in an ecosystem. Our connection to nature is so intense and unmitigated that we are seeing identical plague-like behaviour among mice and grasshoppers, as well as in our own obesity epidemic. By reflecting our own problems in a small mirror and using our understanding of wildlife to address the causes, we might just be able to find a more sustainable future for all humanity.

Redressing the imbalance requires tweaking the load we place on our planet. This is going to happen naturally as we move towards increased organic farming where we'll have more predators, less nitrogen and more diversity. If we can speed that process up by choosing to limit our use of pesticide and fertiliser and rewild one-third of farmland, then it's highly likely that will be sufficient to protect our food, health and climate.

There are those who argue that rewilding, regenerative farming and other like measures are unviable, that food production will suffer and cannot be done that way. To that I would say, wise up. Who says that every piece of land must be entirely profitable? We've established that animals are both destroyers and creators, so there is no such thing as a planet where everyone creates all the time. Providing food and shelter will always destroy ecosystems and all this means is that we need to connect places that are completely 'wild' with places that are semi-wild to maintain a dynamic-but-steady stable state.

One might say that chronic stress is to behaviour what climate change is to ecosystems; our dependence on sugary snacks and ultra-processed foods is like burning fossil fuels. We could choose to turn that off any time. But to restore that underlying autonomy we need to make healthy food choices just as we need to rewild landscapes. This is an essential step to reversing rampant global anxiety caused by our disconnection from nature, because the biggest threat to humanity is our disrespect for wildlife and natural processes.

We know that once we restore nature's balance the most serious existential threats go away: climate warming and extreme weather, malnutrition, obesity and other diseases, drought and nutrient pollution. The lesson we learn from looking at grasshoppers is that stress isn't limited to people. It's become a sweeping, planet-wide epidemic affecting the quality of life for all animals.

As Oswald says:

All animals are trying to do is to fulfil their biological natures. When we look at an insect and think 'let's just kill it' we don't think of it as a living being trying to do the same thing as you and me. If we would view these animals differently, I think we would begin to appreciate these living beings as centres of life.

By allowing wildlife populations a chance to rebuild, they will do the hard work and reconnect us with nature. We need to let go of our fear and allow them to take the lead. This is perhaps the greatest lesson of all.

Will Humans Survive the Next 100 Years?

The same ego that makes us think we can save the planet using our technology is the same one that makes us think we can destroy it, and it's that false sense of superiority that undermines our faith in nature. The truth is that humans will not destroy Earth. A certain inbuilt incompetence restricts our ability to do as much damage as we might imagine. Ironically, it is our fallibility as a species that gives me the greatest hope for humans.

My kids are at an age where they're starting off in the world. They are the first generation to have been bombarded each day with rhetoric designed to make them feel guilty about being human and instil despair about the future. Many people feel strongly about the state of our world and are often made to feel responsible for our species, our evolution and our society's failures. But there are so many factors beyond our control – pandemics, brain parasites, caffeine addiction, poorly elected leaders, profit-driven malnutrition – that we cannot possibly be responsible for. If you say that today, then opponents to conservation might even gaslight you for hypocrisy, but that's plain dumb. Living inside an economy that restricts our autonomy for everyday decisions shows we are very much not in control.

These days if you have any interest in nature, every second you spend online can feel dominated by messages designed to make you feel guilty about it. Peer-reviewed scientists have even started blaming humans for the extinction of woolly mammoths[120], which is absurd. Today's humans can't possibly be responsible for something that happened 120,000 years ago, especially when our ancestors back then were living traditional lives strongly connected to landscapes and the wildlife around them.

Though we are repeatedly told we are having too much of an impact on the natural world, at the same time fellow conservationists can blame you for not doing enough. But no individual can solve anything alone, and clearly whole species can't even solve the world's problems alone.

These days I regularly find myself in conversations with friends and colleagues who have lost hope, though after a robust half hour of conversation I'm heartened to often hear them say they feel better. Everyone has been brainwashed into thinking we're irreversibly doomed, but it's not the case.

We need to regain some sense of perspective by finding out about some of the extraordinary changes that are sweeping over Earth right now, the actions highlighted in the chapters of this book. Some of them are engineered, most are natural and all are driven by nature taking its course. Most things are happening for the better.

Half of the world's animals are increasing in number[2]. These are the seeds of recovery in a system that has begun to put humanity back in its place. In my lifetime I have never seen more commitment to nature conservation than I do today and that is truly motivating. You won't, however, find enough of this on ultra-processed feeds from TV or social media. You need to strike out on your own, re-enter the wilderness and seek a better diet among healthier information sources. Like your uncanny ability to choose the right food, inherent in your DNA is the ability to seek knowledge that matters. Our minds and souls are connected to other animals in a way that we have long been aware of, even if we've buried this knowledge in the back of our minds. All it takes to rediscover this is to step back into nature.

Each generation adapts to survive in its own time and accepts that as normal. Our survival as a species won't come about because of politicians or scientists, it will come about because of the inherent wisdom of the majority of people like you who are just getting on with life. People who have woken up to the knowledge and wisdom that thousands of human generations before us learned and still know.

Children born a century from now will face different challenges, and who is to say what that will mean? It will feel no different to them, as you feel today. If we succeed in creating a new nature-based economy and wildlife continues to bounce back the way it seems to be, then there's a fair chance they'll be happier than you are now.

If we consider the words of the late peace activist Thich Nhat Hanh: 'If someone is not happy within themselves, how can they help the environment? That is why to protect non-human elements is to protect humans, and to protect humans is to protect non-human elements[121].'

In nature, humans and animals have equality, and if anything it is only our ego, pride and self-worth that creates loathing for our own species. We should not feel guilty about being human. We should not feel guilty about following our animal instincts. We should try not to blame anyone for what we can't control and instead work to include them in a way forward. Most of all we need to understand that we are not in complete control of our species' future because forces more powerful than us also determine this. All we can do is try to be better stewards of Earth, transform into a nature-based economy and influence others to do the same.

Indigenous knowledge is a window into our past. It's the ultimate lesson from nature as it comes from our own species of animal that we can talk to in a language of spoken words that we understand. It's not about living primitive lives, rather it is about using the best of our history to have the most sophisticated lives possible and stand up to the challenges we might face in future.

Beyond that are the infinite lessons we can take from wildlife. There is growing discourse and research pointing to how similar we are. Soon we will simply accept that animals and humans share the same fate and that any animal, however rare or insignificant seeming, could be crucial to our survival. A decade ago no-one would have thought the humble and near-extinct eastern barred bandicoot could help bolster farm profits. Now our knowledge is different because we are starting to look for it in the right places and ask better questions. But these aren't revelations because past civilisations knew this. Humanity has been here before.

We must fight to save every animal we can because animals are our medium through which we can talk to the whole of nature. As a naturalist this has been one of the most challenging and glorious conclusions of my work. I'm not denying the importance of research, but I question the motivations of scientists and engineers who seek to control nature before treating animals as 'centres of life', as Oswald says. Or those who have lost touch with the inexplicable yet fundamental connection we have with nature.

I'm not certain we need to do much more discovery. What we need now is science that applies what we learn from wildlife to our greatest environmental challenges. Science needs to rewild itself. It needs to take a leaf from the ecosystem, embrace community wisdom and inspire participation instead of deciding (internally) what should be done. We need to see nature as integral to all our livelihoods and lifestyles and sail on its breeze, not power into it from outside.

Will humans survive? It depends on how long we're talking about. Our modern society is barely even visible on humanity's own timeline, let alone Earth's. If you're reading this in the evening and define the existence of *Homo sapiens* as one year, then the Industrial Revolution happened around lunchtime. Disease-addled Egyptian leaders were being buried with their cats last week. Rome disappeared three days ago. Most modern wildlife decline happened since dinner time. Many of the efforts to rewild the planet that we've discussed here happened since you read the last three pages. By the time you've turned the last page hopefully we'll be well on our way to building a new nature-based economy.

The best hope we have is knowing that we have done enough to save the most important species on Earth because these are key to the resilience of our species. If humans have existed for the equivalent of a year then orangutans evolved 75 years ago and elephants 300 years ago. As a large-bodied mammal living halfway up the food chain we can't afford to lose the giants on whose shoulders civilisation was bult. We cannot survive on a diet of algae and insects because we are hunters, gatherers and farmers, not filter feeders.

We don't have the skills or technology to replicate nature-based processes on a large scale but we can use this to limit our impact while we allow wildlife populations the 20 to 100 years needed to rebuild an abundance large enough to begin rebalancing ecosystems. The best hope will be in the places we have left where wildlife still thrives, where communities wise enough to hold onto their beliefs and protect their environment will continue to survive.

I think of the islanders in Fiji or the First Nations people still living remotely in outback Australia. They may never read this book. They don't have to. They will continue to adapt the way they always have. Their cultures suffered the worst collapse 100 years or more ago under the influence of colonial rule and they've been rebuilding ever since. A future that looks bleak for us now could already be brighter for them. How comfortably we survive the collapse of our culture depends on how quickly we can restore similar beliefs in nature and protect the wildlife around us.

The magic of nature is in the vast array of surprising adaptations and opportunities for recovery. In the millions of years before humans evolved the animals we take for granted had already withstood occasional calamitous changes to Earth's biosphere. Saiga antelope can double their population annually. Some sea urchins, lobsters and naked mole rats never age. Locusts can quadruple their food intake, moving to new pastures to restore grassland ecosystems after periods of famine. Recovery is an inbuilt mechanism that animals do for us. If you know where to look you can see it happening fast, everywhere and in areas we least expect it. The animals we call 'pests' are the best examples of this and we would do well to give them a bit of respect as they are at the frontline of restoring broken life support.

A combination of progressive policies and ecosystem decline have already reopened the chance for wildlife to flourish again. In Europe in the space of 23 years the number of Iberian lynx has risen from 94 to over 2,000. Blue wildebeest in the Serengeti were reduced to 300,000 leading to increased vegetation and wildfires turning the landscape into a source of atmospheric carbon. Their numbers have since increased to 1.5 million and the grasslands are a carbon sink again. Northern hairy-nosed wombats have increased from a historic low of 60 in 1991 to over 350 today. There are countless other examples. All in all, two-thirds of conservation actions have been proven to work wonders[122], and investment in conservation has more than tripled since 2012[123]. It seems we are living in an enlightened age when people are quickly relearning the value of nature.

The effort we make in conservation science isn't wasted, and maybe even our failures aren't a reason to feel depressed because, after all, it is failure that is the mother of invention. The most exciting thing of all is the change that's happening all around us that we're too preoccupied to notice. All over the world animals are making themselves more visible and present in our daily lives. As humans and wildlife are forced together to face the same challenges of degraded ecosystems, we are becoming noticeably more co-dependent. Animals are tolerating human presence where in the past they shied away because they were hunted. Tourism is partly responsible for this. It may not be a long-term economic solution for conservation – and isn't always done sustainably – however it is extraordinary to see people from different cultures coming together and caring for wildlife through shared appreciation and wonder.

Fear, anxiety, depression and helplessness come from a lack of connection to nature and purpose. Biruté Galdikas is one among the very few humans I know who has spent long enough in the company of animals to have given this proper thought. For 50 years Biruté lived entirely in the orangutans' universe, gaining a sense of perspective and inner peace through understanding the intricate complexities of the life, trials, tribulations and resilience of living creatures including people and other animals.

This is evidence that our sense of place on Earth can't be learned from history books or research; it comes from a deeper, more abstract place that cannot be fully explained in words. It is a spiritual wisdom that was traditionally passed on from person to person, generation to generation in music, dance and stories. This is how a lifetime of observing nature for one person becomes a gateway to the rest of us understanding the significance of seemingly benign interactions we might have with wildlife throughout our daily life.

The life's work of Biruté, Jane, Dian and others like them has already inspired generations to change their attitudes to wildlife. Unknowingly, they may have been at the forefront of a momentous change in recent human evolution where we are relearning to commune with other animals. The fascination we have with nature, the connections we seek every day with wild animals and the animals we keep as pets is a survival instinct that permeates every cell in our body. It should come as no surprise, therefore, that worldwide publicity of wildlife encounters through the likes of *National Geographic* and the BBC's *Natural History Unit* has led to billions of people becoming fascinated by animals.

This massive shift in humanity's behaviour around wildlife, our actions and the reactions of wild animals to a changing landscape has completely altered the relationship we have with each other. 'What I find interesting is that some people are habituating orangutans now and they seem to find it simpler,' says Biruté. 'I think it's because the nature of the ecological and environmental situation on this planet has changed.'

In their paper 'Orangutans venture out of the rainforest and into the Anthropocene', Stephanie Spehar and her co-authors tell us how for 70,000 years humans have shaped the destiny of orangutans[74]. Of course, the authors didn't write about how much impact the orangutans had on people, and that's why this book you're reading right now is important. The authors found that while the orangutans are vulnerable to hunting and deforestation, they are more tolerant to other human activities. Of course they are! Otherwise their kind could never have survived. In other words, remove the main threats from us and we can once again enjoy a mutually beneficial relationship.

We should all sit down for a while and ponder this notion. We're so engrossed in our own self-indulgent politics that are largely disconnected from the reality of ecosystems that we've invented the idea we are all doomed. We haven't stopped to consider for a moment that nature's support might be about to improve enough that we really can build a better future.

Earth's biosystems are brutally inflexible though, and we are coerced into playing by the rules. It's the adaptive behaviour of animals that permits some elasticity within the confines of our planet's physics. While no-one can tell the future, we now know for a fact that the human population cannot possibly survive at the level it is today.

Grasshoppers have taught us that global obesity is a last-ditch survival strategy for any animal population under stress, that it is a natural response to fear and famine. We simply can't sustain ourselves by cramming more sugars into our bodies and seeking new foraging grounds because there is no greener grass left – we've exhausted our options. This outcome is beyond our control, so the only possible way forward is for our current population to decline while we rebuild what we've lost. Surviving cultures and customs that are fit for purpose will be those that simultaneously maintain a balanced diet and a healthy ecosystem. Overall life spans will decrease, but our reproduction rate will stabilise to maintain fewer people on Earth.

What our population ends up being depends not on our technology but on how many animals are left around us to provide the nutrients we need. Over the last couple of hundred years we have taken a finite amount of energy inside the biosphere and we have borrowed this from the Earth. We have stolen so much from wildlife that we are undermining our own potential. Climate change is partly a symptom of that decline. If we had not killed so much wildlife, we might not already be facing the 'crisis' that fossil fuel burning imposes on us today.

We should be concerned about our rampant creation of forever chemicals. The universal use of soft plastic may also impact us heavily in the coming few hundred or thousands of years. It's already thought this is why male fertility has declined by half since 1973. Add to that the continued bioaccumulation and ingestion of poisons that cover our food systems, from rodenticides to glyphosate. Scientists still argue over what level of consumption is individually safe but that's not the point. Have we learned nothing from nature? It has never been about what we put inside ourselves individually, it's the risk of to the ecosystems we change outside ourselves that matters most. We often ask the wrong questions and remain blissfully ignorant of the most likely triggers of our civilisation's decline.

But it's good news overall. If we can create a nature-aware economy, the debt we have to the Earth can be repaid very quickly with the help of animals. Wild animal populations will continue to rebound. These may be some of the animals we call pests but they are among the most superbly adept creatures to have ever roamed the Earth. The highly stressed plagues of fauna like locusts, cockroaches, mice, deer, foxes, rabbits, sea urchins, feral cats and dogs will even settle down and become less inconvenient.

The energy we once consumed, when wasted and poured into our oceans and atmosphere, will be rapidly consumed by many animals and within decades new food chains and landscapes will form. In a relatively short time we will be living a different existence, and our culture will once again reform itself and change to fit the new fashion.

Fifty years after we are predicted to reach peak population (around 2090) we will start to see a reversal of the losses that biodiversity collapse has caused in the last 50 years. These are the losses we didn't realise (until now) could be even more significant than climate change. That's good news. Without our constant interference and erosive threats, animal life-support mechanisms will bounce back even more quickly and become able to adjust faster to fluctuations in climate. This buys us extra time to do the things we need to do – to live more sustainably. By that time our world will certainly be very different to how it is today; it will inevitably be a world richer in wildlife.

If you're left wondering what to do, fear not. There are monumental changes happening and much to be positive about. Nature, as we have discovered, is incredibly supportive but only once we fix our relationship with it. In the next few years you will begin to experience changes in the way we live. Some – like electric cars – will be obvious, while others will emerge as a way of life that we take for granted.

The fact remains that most of us cannot drop everything and 'do something' on a grand scale for the planet. But as individuals we can support the emergence of more sustainable practice, enable change and behave better in our daily lives. We can't all be environmental scientists, as we still need doctors, teachers, nurses and baristas. We want to be able to live the lives we choose and support our community, all while knowing that nature can once more look after us long-term.

There is no need to despair, be stressed or feel anxious or guilty. If you haven't yet chosen the path of learning from nature then get outside and start discovering new lessons for yourself. Switch off your worries and turn on your instincts. Nature's plan is for you and your descendants to embrace the opportunities it gives us to survive. Other less fitting behaviours simply die away. The next exciting chapter in our story of humanity will be about how society is transforming itself to avoid our own worst excess. It's a path we are on already and one we all need to support to pave the way to a brighter and more comfortable future.

Armed with this knowledge and surrounded by a renewed abundance of wildlife, humans can not only survive but thrive.

Acknowledgements

Thank you to everyone who made this book possible. To my parents William and Barbara Mustoe, Carla, Sadie, Charlie and all my friends, especially Gavin and Jess Burbidge for their constant advice and for a valuable translation of information from French. Thank you to Dennis Jones, Debbie McInnes, Jen Bowden and Nada Bakovic for keeping me honest and doing such a wonderful publishing job for me. A big thank you to everyone who I interviewed for the book, most notably Dinah Nieburg, Anissa Lawrence, Oswald Schmitz and Dr Biruté Mary Galdikas, as well as the very many authors, scientists, journalists, film-makers, naturalists and others whose work is listed in the references. A particular thank you to Fred Smith for allowing me to reproduce the poem 'Sparrows of Kabul' in the chapter of the same name. Finally, thank you to anyone and everyone who understands the importance of wildlife and doesn't need to be persuaded that animals are humanity's best hope.

References

1. Haddaway, N. and D. Leclère. *WWF Living Planet Report 2020*. 2020.

2. Toszogyova, A., J. Smyčka and D. Storch. 'Mathematical biases in the calculation of the Living Planet Index lead to overestimation of vertebrate population decline.' *Nature Communications*. 15(1), p. 5295. 2024.

3. Vynne, C., et al. 'An ecoregion-based approach to restoring the world's intact large mammal assemblages.' *Ecography*. 2022(4). 2022.

4. Chapron, G., et al. 'Recovery of large carnivores in Europe's modern human-dominated landscapes.' *Science*. 346(6216), p. 1517–1519. 2014.

5. Carroll, Sean B. 'How we can revive Planet Earth in less than 20 years.' *Big Think*. YouTube. 2024.

6. Ritchie, H. 'Wild mammals are making a comeback in Europe thanks to conservation efforts.' *OurWorldInData.org*. 2022. https://ourworldindata.org/europe-mammal-comeback

7. Rosling, H and O. 'How not to be ignorant about the world'. TED. YouTube. 2024.

8. Woinarski, J.C.Z., S.T. Garnett and K.K. Zander. 'Social valuation of biodiversity relative to other types of assets at risk in wildfire.' *Conservation Biology*. 38(3), p. e14230. 2024.

9. Gammage, B. *The Biggest Estate on Earth: How Aborigines made Australia*. Allen & Unwin. 2011.

10. Wettenhall, G. and Gunditjmara people. *The People of Budj Bim: Engineers of aquaculture, builders of stone house settlements and warriors defending country*. Second edition. EM Press. 2022.

11. McNiven, I.J. and D. Bell. 'Fishers and farmers: Historicising the Gunditjmara freshwater fishery, western Victoria'. *The Latrobe Journal*. 85. 2010.

12. Bernthal, F., et al. 'Nutrient limitation in Atlantic salmon rivers and streams: Causes, consequences, and management strategies.' *Aquatic Conservation Marine and Freshwater Ecosystems.* 32, p. 1073–1091. 2022.

13. Doughty, C., et al. 'Global nutrient transport in a world of giants.' *Proceedings of the National Academy of Sciences of the United States of America.* 113. 2015.

14. Bellamy, A.R. and J.E. Bauer. 'Nutritional support of inland aquatic food webs by aged carbon and organic matter.' *Limnology and Oceanography Letters.* 2(5), p. 131–149. 2017.

15. Carew, K. *Beastly: The Epic 40,000-Year Story of Animals and Us.* Canongate Books. 2024.

16. Gies, E. *Water Always Wins: Thriving in an Age of Drought and Deluge.* University of Chicago Press. 2022.

17. de Kock, W., et al. 'Threatened North African seagrass meadows have supported green turtle populations for millennia.' *Proceedings of the National Academy of Sciences of the United States of America.* 120, p. e2220747120. 2023.

18. National Geographic. 'Key Components of Civilization.' *National Geographic Education.* Accessed 2024.

19. Barber, G. 'The Trillion-Dollar Auction to Save the World.' *Wired Magazine.* 2023.

20. Waycott, M., et al. 'Accelerating loss of seagrasses across the globe threatens coastal ecosystems.' *Proceedings of the National Academy of Sciences for the United States of America.* 106(30), p. 12377–12381. 2009.

21. Ng, K. 'Alan Titchmarsh warns against "ill-considered" rewilding trend in domestic gardens.' *The Independent.* 2023.

22. Jesmer, B.R., et al. 'Is ungulate migration culturally transmitted? Evidence of social learning from translocated animals.' *Science.* 361(6406), p. 1023. 2018.

23. Ohlson, K. *Sweet in Tooth and Claw: Nature is more cooperative than we think.* Scribe. 2022.

24. Gilbert, N.A., et al. 'Human disturbance compresses the spatiotemporal niche.' *Proceedings of the National Academy of Sciences for the United States of America.* 119(52), p. e2206339119. 2022.

25. Meyer, C.J., et al. 'Parasitic infection increases risk-taking in a social, intermediate host carnivore.' *Communications Biology*. 5(1), p. 1180. 2022.

26. Thomas, B. 'Meet the Parasites That Control Human Brains.' *Discover Magazine*. 2020.

27. Johnson, S.K., et al. 'Risky business: Linking *Toxoplasma gondii* infection and entrepreneurship behaviours across individuals and countries.' *Proceedings of the Royal Society B: Biological Sciences*. 285(1883), p. 20180822. 2018.

28. Flegr, J., et al. 'Toxoplasmosis – A Global Threat. Correlation of Latent Toxoplasmosis with Specific Disease Burden in a Set of 88 Countries.' *PloS one*. 9, p. e90203. 2014.

29. Kopecky, R., et al. *Le Petit Machiavellian Prince: Effects of latent toxoplasmosis on political beliefs and values*. 2021.

30. Ellwood, B. 'A common parasitic disease called toxoplasmosis might alter a person's political beliefs.' *PsyPost: Evolutionary Psychology, Mental Health, Political Psychology, Racism and Discrimination*. 2022

31. Gondolo, T. *Fifty Shades of Gray Matter*. Atmosphere Press. 2023.

32. Zhou, A. and E. Hyppönen. 'Long-term coffee consumption, caffeine metabolism genetics, and risk of cardiovascular disease: a prospective analysis of up to 347,077 individuals and 8368 cases.' *The American Journal of Clinical Nutrition*. 109(3), p. 509–516. 2019.

33. de Balzac, H. *Traité des excitants modernes*. 1832. [Translated with kind assistance from Jess Burbidge]

34. Pollan, M. *This Is Your Mind on Plants*. Penguin Publishing Group. 2021.

35. Galdikas, B.M.F. *Reflections of Eden: My Life with the Orangutans of Borneo*. Indigo. 1996.

36. Smith, F. *The Sparrows of Kabul*. Puncher & Wattmann. 2023.

37. Parker, E. 'Climate and Australia's National Security.' *The Forge*. 2022.

38. NOFF. *The Toxic Truth*. Neighbours of Fish Farming: NOFF Short Films. 2022.

39. Porter, J.R. 'Rising temperatures are likely to reduce crop yields.' *Nature*. 436(7048), p. 174. 2005.

40. Strona, G. and C.J.A. Bradshaw. 'Co-extinctions annihilate planetary life during extreme environmental change.' *Scientific Reports*. 8(1), p. 16724. 2018.

41. Ding, H., et al. *Roots of Prosperity. The Economics and Finance of Restoring Land*. 2017.

42. Burgess, M.G. and S.D. Gaines. 'The scale of life and its lessons for humanity.' *Proceedings of the National Academy of Sciences for the United States of America*. 115(25), p. 6328. 2018.

43. Ariando, W. *Developing A Model for the Integration of Bajau Traditional Ecological Knowledge in the Management of Locally Managed Marine Area: A case study of Wakatobi Regency, Indonesia*. 2022.

44. Sullivan, H. 'A sea urchin: they are method actors performing *The Waste Land*.' *The Guardian*. 2023.

45. Chomicki, C. 'Popular seafood species could help curb coral-eating starfish pest, say scientists.' *ABC News*. 2021.

46. Webb, M., et al. 'Chemosensory behaviour of juvenile crown-of-thorns sea star (Acanthaster sp.), attraction to algal and coral food and avoidance of adult conspecifics.' *Proceedings of the Royal Society B: Biological Sciences*. 291(2023), p. 20240623. 2024.

47. Day, J., et al. 'Investigating the diets and condition of *Centrostephanus rodgersii* (long-spined urchin) in barrens and macroalgae habitats in south-eastern Australia.' *Marine Ecology Progress Series*. 729. 2023.

48. Stacey, N., et al. *Assessing Traditional Ecological Knowledge of Whale Sharks (Rhincodon typus) in eastern Indonesia: A pilot study with fishing communities in Nusa Tenggara Timur*. Report prepared for the Department of the Environment, Water, Heritage and the Arts, Canberra. 2008.

49. Szabo, K. 'Terrestrial hermit crabs (*Anomura: Coenobitidae*) as taphonomic agents in circum-tropical coastal sites.' *Journal of Archaeological Science*. 39. 2012.

50. Starling, S. 'Blue murder: Why NSW's blue groper ban hurts recreational fishers.' *Boatsales.com.au*. 2024.

51. Franzitta, G. and L. Airoldi. 'Fish assemblages associated with coastal defence structures: Does the surrounding habitat matter?' *Regional Studies in Marine Science*. 31, p. 100743. 2019.

52. Sherman, K., et al. 'Spatial and Temporal Variability in Parrotfish Assemblages on Bahamian Coral Reefs.' *Diversity*. 14, p. 625. 2022.

53. Schelske, O., et al. *Biodiversity and Ecosystem Services: A business case for re/insurance.* 2020.

54. Swiss Re. *A fifth of countries worldwide at risk from ecosystem collapse as biodiversity declines, reveals pioneering Swiss Re index.* Swiss Re Group. 2020.

55. Liao, Y. 'Home Insurance in Crisis.' *Milken Institute Review.* Milken Institute. 2024.

56. Ferrario, F., et al. 'The effectiveness of coral reefs for coastal hazard risk reduction and adaptation.' *Nature Communications.* 5(1), p. 3794. 2014.

57. Bellwood, D., A. Hoey, and J. Choat. 'Limited functional redundancy in high diversity systems: Resilience and ecosystem function on coral reefs.' *Ecology Letters.* 6, p. 281–285. 2003.

58. Adger, W., N. Arnell and E. Tompkins. 'Successful Adaptation to Climate Change Across Scales.' *gec.* 15, p. 77. 2005.

59. Steneck, R.S. and C. Johnson. 'Kelp forests: Dynamic patterns, processes, and feedbacks.' *Marine Community Ecology and Conservation.* p. 315–336. 2013.

60. Baines, G. 'Agriculture, biodiversity conservation and protected areas.' *Lessons from Global Experience.* International Centre for Environmental Management. 2003.

61. Global Rewilding Alliance. 'Rewilded Bison are climate heroes – new research.' Global Rewilding Alliance with Yale University. 2024.

62. Orwell, G. *Animal Farm.* HarperCollins Canada. 1945.

63. IFOA. *Impacts of Biodiversity Loss Part 1: Economic impacts.* Institute and Faculty of Actuaries: IFoA's Biodiversity Working Party. 2023.

64. European Commission. *Nature Restoration Law: For people, climate, and planet.* EUGreenDeal. 2022.

65. UNEP. 'Global annual finance flows of $7 trillion fuelling climate, biodiversity, and land degradation crises.' UN Environment Programme. 2023.

66. Singleton, G., et al., eds. 'Ecologically-based rodent management.' *ACIAR Monograph No. 59*, p 494. Australian Centre for International Agricultural Research. 1999.

67. Shine, R. 'Stop killing brown snakes – they could be a farmer's best friend.' *The Conversation.* 2024.

68.   Bianchi, F.J.J.A., et al. 'Spatial variability in ecosystem services: simple rules for predator-mediated pest suppression.' *Ecological Applications*. 20(8), p. 2322–2333. 2010.

69.   Ecocide Law. *Legal Definition and Commentary 2021*. Ecocide Law website. 2021.

70.   Johnston, G. and C. Menz. *An independent review of the evidence underpinning the rewilding of southern Yorke Peninsula*. 2019.

71.   DePalma, E. *How canopy structure affects orangutan nesting sites*. The Orangutan Conservancy. 2019.

72.   Redmond, I. 'Primates, Biodiversity and Climate.' *Primate Eye*. The Primate Society of Great Britain. 2021.

73.   Simon, D., G. Davies and M. Ancrenaz. 'Changes to Sabah's orangutan population in recent times: 2002–2017.' *PLoS One*. 14(7), p. e0218819. 2019.

74.   Spehar, S.N., et al. 'Orangutans venture out of the rainforest and into the Anthropocene.' *Science Advances*. 4(6), p. e1701422. 2018.

75.   Pye, S. *Saving Sun Bears: One Man's Quest to Save a Species*. Estralita Publishing. 2021.

76.   Augeri, D. *On the biogeographic ecology of the Malayan sun bear*. 2005.

77.   Yunkaporta, T. *Sand Talk: How Indigenous Thinking Can Save the World*. Text Publishing. 2019.

78.   Martin, S.A. *The Aardvark as an ecological engineer in the Eastern Karoo: Dig patterns and emergent processes*. Nelson Mandela Metropolitan University. 2017.

79.   Pandian, N. 'Beyond "elephant whispers" we need to hear the Kattunayakar's voice.' *The News Minute*. 2023.

80.   ISEC. 'ISEC Nepal: Human-elephant conflict around Bardia National Park.' International Student Environmental Coalition. 2021.

81.   Geertz, T. 'For a peaceful coexistence of people and elephants in Nepal.' Global Nature Fund.

82.   Wohlleben, P. *The Power of Trees: How Ancient Forests Can Save Us If We Let Them*. Schwartz Books. 2023.

83.   Herkenrath, T. 'Hungry herbivores and thirsty plants: How does wildlife shape tree transpiration in a Namibian savanna?' *Functional Ecologists*. 2024.

84.   Kolbert, E. 'Why Is the Sea So Hot?' *The New Yorker*. 2024.

85.   Munzel, T., et al. 'Cardiovascular Effects of Environmental Noise Exposure.' *European Heart Journal*. 35. 2014.

86.   Pizarro, A. 'Una torre-nido única en España combate el virus del Nilo en Coria del Río/A unique nest tower in Spain combats the Nile virus in Coria del Río.' *Diario de Sevilla*. 2023.

87.   Falkenberg, G., et al. 'Avian Magnetoreception: Elaborate Iron Mineral Containing Dendrites in the Upper Beak Seem to Be a Common Feature of Birds.' *PLoS One*. 5(2), p. e9231. 2010.

88.   Schulz, K. 'Why Animals Don't Get Lost.' *The New Yorker*. 2021.

89.   Schwartz, J.D. *The Reindeer Chronicles: And other inspiring stories of working with nature to heal the Earth*. Chelsea Green Publishing. 2020.

90.   Paterson, E. 'Landmark reports reveal UK seabed carbon storage potential.' *Scottish Association for Marine Science*. 2024.

91.   Thomson, A. 'New data reveals how sea mud is far more important than we think.' *Channel 4 News*. 2024.

92.   Chen, X.D., et al. 'Stabilizing Effects of Bacterial Biofilms: EPS Penetration and Redistribution of Bed Stability Down the Sediment Profile.' *Journal of Geophysical Research: Biogeosciences*. 122(12), p. 3113–3125. 2017.

93.   Nyffeler, M. and K. Birkhofer. 'An estimated 400–800 million tons of prey are annually killed by the global spider community.' *The Science of Nature*. 104(3), p. 30. 2017.

94.   Guiden, P., et al. 'Effects of management outweigh effects of plant diversity on restored animal communities in tallgrass prairies.' *Proceedings of the National Academy of Sciences for the United States of America*. 118, p. e2015421118. 2021.

95.   Robinson, J.M., et al. 'Restoring soil biodiversity.' *Current Biology*. 34(9), p. R393–R398. 2024.

96.   Wilson, J.B., et al. 'Plant species richness: the world records.' *Journal of Vegetation Science*. 23(4), p. 796–802. 2012.

97.   Cook, M., J. Douglass, and D. Mallon. 'The economic impact of allergic disease in Australia: not to be sneezed at.' *Aust Soc Clin Immunol Allergy (ASCIA)*. p. 1–111. 2007.

98.   Lakeram, S., et al. 'Coprolites in Middle Pennsylvanian Cordaitean Cones: Evidence for Early Pollinovory in Cordaiteans?' 2019.

99. Linskens, H.F. 'Mature Pollen and its Impact on Plant and Man in Sexual Plant Reproduction.' Springer Berlin Heidelberg. p. 203–217. 1992.

100. Gassner, M., R. Gehrig, and P. Schmid-Grendelmeier. 'Alder pollen of *Alnus spaethii* at Christmas: From epidemiology of molecular allergens to the political solution.' *Clinical and Translational Allergy.* 4(2), p. P36. 2014.

101. Miller, S.G. *Guano and the Opening, Pacific World: A global ecological history.* Cambridge University Press. 2010.

102. Blackall, T., et al. 'Ammonia emissions from seabird colonies.' *Geophysical Research Letters.* 341. 2007.

103. Croft, B., et al. 'Contribution of Arctic seabird-colony ammonia to atmospheric particles and cloud-albedo radiative effect.' *Nature Communications.* 7(1), p. 13444. 2016.

104. Wilson, L.J., et al. 'Modelling the spatial distribution of ammonia emissions from seabirds in the UK.' *Environmental Pollution.* 131(2), p. 173–185. 2004.

105. Katz, C. 'An Icelandic Town Goes All Out to Save Baby Puffins.' *Smithsonian Magazine.* 2023.

106. Dodson, S., et al. 'Long-distance communication can enable collective migration in a dynamic seascape.' *Scientific Reports.* 14(1), p. 14857. 2024.

107. Angela, A., et al. *IUCN Global Standard for Nature-based Solutions.* IUCN. 2020.

108. United Nations. 'COP15 ends with landmark biodiversity agreement.' United Nations Environment Program. 2022.

109. Yallop, C. *Macquarie Dictionary.* 2005.

110. Goeckeritz, I. *The Rights of Nature: A Global Movement.* YouTube. 2021.

111. Dahlstrom, M. '152 koalas killed in private forest by US company: "Not good enough".' *Yahoo News.* 2023.

112. Alcoa. 'Portland Aluminium establishes new koala habitat.' Alcoa Australia. 2023.

113. Clover, C. *Rewilding the Sea: How to Save our Oceans.* Ebury Publishing. 2022.

114. Ocean Earth Foundation. 'SeaNet Indonesia: Sustainable Livelihood for Arafura Communities.' 2017.

115. Keighley, M. and T. Huth. *Tales of an Outback Guide, or Why Kangaroos go Boing, Boing, Boing.* Wildiaries Publishing. 2012.

116. Mustoe, S. and N. Hayward. 'How a Grasshopper and a Farmer Changed the Face of Arnhem Land'. *Wildiaries.* YouTube. 2014.

117. Raubenheimer, P.D. and P.S.J. Simpson. *Eat Like the Animals: What Nature Teaches Us about the Science of Healthy Eating.* HarperCollins Australia. 2020.

118. Victoria State Government. 'Heart disease and mental health.' The Victorian State Government: Better Health Channel.

119. Segerstrom, S. 'Stress Affects Immunity in Ways Related to Stress Type and Duration, as Shown by Nearly 300 Studies.' American Physiological Association: Press Release. 2004.

120. Lemoine, R.T., R. Buitenwerf and J.C. Svenning. 'Megafauna extinctions in the late-quaternary are linked to human range expansion, not climate change.' *Anthropocene.* 44, p. 100403. 2023.

121. Hanh, T.N. *Zen and the Art of Saving the Planet.* Ebury Publishing. 2021.

122. Langhammer, P., et al. 'The positive impact of conservation action.' *Science.* 384, p. 453–458. 2024.

123. The Nature Conservancy Australia. 'Closing the Nature Funding Gap: A Finance Plan for the Planet.' The Nature Conservancy Australia. 2020.

Index

ABOUT THE AUTHOR

Simon Mustoe is an ecologist, artist, expedition leader and naturalist with a passion for connecting humans with the natural world. He's tumbled in boats amid frigid North Atlantic storms, trekked solo into Madagascar's remote dry forests and sailed the archipelagos of West Papua. As a teenager he helped produce BBC nature documentaries and as an adult he has worked with some of the world's leading conservation, wildlife and ecology organisations. His first book, *Wildlife in the Balance*, received critical acclaim from the likes of Dame Joanna Lumley and Ian Redmond OBE. He lives in Melbourne, Australia.

www.ingramcontent.com/pod-product-compliance
Lightning Source LLC
Chambersburg PA
CBHW032044040426
42334CB00039B/1131

9780645453584